Stephen Martin Saxby

Saxby's Weather System

Or, Lunar influence on weather

Stephen Martin Saxby

Saxby's Weather System
Or, Lunar influence on weather

ISBN/EAN: 9783337034900

Printed in Europe, USA, Canada, Australia, Japan

Cover: Foto ©berggeist007 / pixelio.de

More available books at **www.hansebooks.com**

SAXBY'S WEATHER SYSTEM

OR

LUNAR INFLUENCE ON WEATHER.

BY

S. M. SAXBY, Esq., R.N.

PRINCIPAL INSTRUCTOR OF NAVAL ENGINEERS H. M. STEAM RESERVE;
LATE OF GONVILLE AND CAIUS COLLEGE, CAMBRIDGE:
AUTHOR OF 'CALCULATION AND PROJECTION OF THE SPHERE,' 'THE STUDY OF
STEAM AND THE MARINE ENGINE,' ETC.

SECOND EDITION.

LONDON:
LONGMAN, GREEN, LONGMAN, ROBERTS, & GREEN.
1864.

PREFACE.

THE recent publication by another person in England of a pamphlet called '*The* Foretelling *of the* Weather,' renders it necessary to substitute a new Title to this Second Edition of 'Foretelling Weather.'

It has been almost all re-written in order to bring out the subject of Predicting Weather to the latest period.

It is nothing singular that, in introducing particulars of a new discovery in Science, opposition should have arisen; but, in this case, the only manner in which to meet adverse opinions is to produce such a mass of facts from strangers who have been totally independent observers, as to place before the general public materials from which any one of ordinary intelligence can draw safe conclusions.

Very many gentlemen, whose kindness in offering and sending assistance by detailing careful and long-continued observations at distant places deserve mention, will, it is hoped, accept unfeigned thanks; but their number is really too great for insertion with propriety. The author would, however, ask permission to make an exception in specially thanking a gentleman of Dover, R. C. DARWALL, Esq., for the very great assistance he rendered at an early period of the investigation, which has led to a great public result—not merely as an advantage to the present generation, but perhaps to all posterity.

CONTENTS.

CHAPTER I.
The Mode and Tone in which my Lunar Theory was first publicly announced . . . . PAGE 1

CHAPTER II.
Its Reception — The 'Weather Book' — Dové's 'Law of Storms' — Astro-Meteorologists — Opinions of eminent Men — Public Interest in the Question of Weather — Facts or Fallacies 12

CHAPTER III.
The World's disbelief in the Moon's Influence on Weather as shown by the 'Weather Book' — 'Luni-Solar' Theory 17

CHAPTER IV.
Sir John Herschel's Opinion is that of all Philosophers — My Claim to the Discovery of the Lunar Theory of Weather Anticipations . . 24

CHAPTER V.
Cause of popular Error explained — The Ancients as Weather-wise as ourselves — Proofs — Progress (?) during 1800 years . . . 29

CHAPTER VI.
The Barometer and Professor Buys-Ballot — Defence of Admiral Fitz Roy's Office and Department — Appeal in behalf of it — Aneroids as Rewards for saving Life &c. — Defence of Telegraph Warnings — Weather Warnings — Shipwrecked Mariners' Society and National Life-Boat Institute 36

CHAPTER VII.

Professor Dovó — Mr. Belville — Mr. Hartnup — Causes of Differences of Opinions explained — Mr. C. Chambers — Evidence of Moon's Influence substantiated . . . . PAGE 46

CHAPTER VIII.

Sir F. Leopold M'Clintock — A Pilot's Testimony — Construction of the Barometer, and its Action 66

CHAPTER IX.

Plymouth Merchants — Scotch Seamen — Somersetshire Agriculturists — Independent Public Testimonies 74

CHAPTER X.

The 10th to 13th December, 1863 — Subsequent Lunar Periods . 82

CHAPTER XI.

Saxby's Weather System explained — Cyclones and their connection with Lunar Influences — Remarkable Cyclone of November 23, 1863, illustrated and described 100

CHAPTER XII.

Fulfilments in 1862 and 1863 — Great Day Auroral Storm of January 1863 — Weather Warnings up to January 1, 1866 . . . 111

SAXBY'S WEATHER SYSTEM.

CHAPTER I.

THE MODE AND TONE IN WHICH MY LUNAR THEORY WAS FIRST PUBLICLY ANNOUNCED.

THE untoward seasons of 1859 and 1860 originated speculations on the weather, which received additional attention from the endeavours of the Board of Trade to turn them to public benefit. The tempestuous period referred to, resulting in losses upon the coast of so many of our seamen, and a large amount of property, affected also the agricultural interests so widely (in the destruction of crops and the hopes and expectations of the farmer), that 'weather' has become an inexhaustible topic with the whole community.

Even yet, as we dash by rail through the festive valleys of our fruitful island, the weather forms the staple burden of remarks, and continued speculations on its probable changes or continuance have proverbial sway over the movements of the multitude; indeed, to consult the meteorological reports and 'forecasts' of the newspapers has become a habit as prevalent as the prying into the state of the funds was with our forefathers. Nor can we touch upon the popular subjects of the day without drifting quickly into a comment upon the season and its probable weather, and the reliance to be placed on this or that 'prognosticator.' Thus in our travels, the very commonplaceness of the subject is a source of much apparently innocent loquacity.

Such then is the subject of weather to Englishmen in general, but there is also a large portion of the community who have a deeper interest in the question, whose very lives are placed in jeopardy by such speculations, and the happiness of whose whole families during their entire earthly career, is involved in this apparently commonplace topic of weather. It is true there exist societies of good men ready, as far as human means and sympathies *can* soothe, to help the widow and orphans of those victims who are yearly sacrificed on the altar of prejudice on the subject of weather. It is prejudice which exhausts the resources of the National Lifeboat Institution, and the funds of the charitable and excellent Shipwrecked Mariner's Society; and its peculiar nature is such that a remedy has in each succeeding year been sought for with increased anxiety, but in vain.

A still larger class of people, whether they be merely pleasure-seekers, or those in pursuit of their out-door avocations, or those whose misfortune it is to be invalids, have also a strong claim on men of science who profess to be learned in meteorology.

With all respect for the illustrious men whose wisdom in other subjects has immortalised their names, their prejudices on the subject of weather seem to have been bequeathed by them to posterity, and are even yet being upheld by living philosophers on whose shoulders certain mantles have fallen. Their decree had gone forth that *the moon does not affect the weather*. To this then we may attribute the strong prejudices which would still hold our judgments in bondage. But surely it is time to free ourselves from every trammel of false philosophy, and look the question in the face, boldly and with independence;

provided a sufficient base is available on which to found a more consistent theorem. Why should it be allowed for one moment that science bars the door against increase of knowledge respecting atmospheric disturbances, while she opens it widely with a welcome to those who search into any other subject?

Hitherto the public thirst to possess a knowledge of coming weather has tempted many an impostor to vend his 'prognostics,' the constant failures in which still further prejudice the whole question; hence the difficulty of obtaining a hearing for any suggestions founded on really honest deductions. The weight of opinions delivered by men of highest calibre in science may well have daunted the attempts and suppressed the convictions of many a casual enquirer. Now, as I am the first who, claiming to be allowed the weight of long experience in mathematical and physical science, has ventured to contest the point, not, however, in a hostile spirit, but in order if possible to elicit the truth, it is for me to set forth honestly and unflinchingly the present state of the question of weather in a manner to be fairly understood by the public; not because there is a theory of weather with which my name has become associated, but because it will be well to show that I have at least consistency behind which to entrench myself. If in so doing I am driven to the unpleasant necessity of arraigning the opinions of others, it is no more than every one who has the good or bad fortune to detect a popular error must submit to. I have dealt in no mysterious *charlatanerie*; reasons for all my opinions have from the first been freely given to the public, and in such plain untechnical language as befitted a question intended for practical advantage to the humblest mariner and the poorest fisherman. I may further, with some self-approval, declare that I have not sought to turn my discovery *to any pecuniary advantage whatever*. I *therefore* touch the subject with 'clean hands,' and, indebted to no mortal for the observations which have led to my discovery of a lunar influence on weather, have facts of my own registration through years (often made by night as well as by day), on which alone I might have based the revised opinions that I have now the honour of submitting to the heads of science and the general public.

That my supposed discoveries were received at first with coldness need not surprise; but recent occurrences seem to have left me once more warm and vigorous in the determined pursuit of this truth, believing, *as I firmly do*, that so desirable a gift to mankind as a fairly developed weather system is within our grasp.

Those who read the 'Nautical Magazine,' which is always open to suggestions for improvement in the condition and welfare of sailors and nautical science, as encouraged by its talented and accomplished official editor, will have seen that for upwards of three years I therein was kindly permitted to challenge scrutiny in connection with my then unfledged theory. Predictions were published months in advance, and I undertook to furnish, from my publicly exposed daily register, details of fulfilments of them—and did so. As an example:—

In September 1861 I had announced, among other predictions, that November 2nd to 5th would probably be a period of *unusual* atmospheric disturbance, and in October I sent printed bills, along the coast especially, warning all against the 14th November, as a period at which a violent cyclone (or hurricane) would most probably pass over Great Britain from the southward and westward. (Long will be remembered the heavy gale of the 2nd November.) And happening to be in London on the *day before* the 14th, and being desirous of offering a public proof of the sincerity of my convictions, I, on the 13th, purposely entered the underwriting room at Lloyd's, to repeat in person my previously announced warning as to the cyclone 'due' on the following day (14th). While explaining to some of the underwriters, one of them remarked that the barometer had already fallen a tenth of an inch since 9 A.M., and even while I was engaged in explaining my reasons for expecting a cyclone, the mercury was still falling; and through the day considerable excitement arose as to my warning, because there were two facts which to the multitude seemed adverse to the probability of any cyclone happening on the day following; for, in the first place, the Board of Trade had informed them that we should most probably have the weather 'moderate till Friday (15th),' and secondly, there was a steady fresh wind from ENE. with cold temperature, while cyclones nearly always approach us with the wind S. to W. However, on the evening of the 13th, unmistakable evidences of the coming hurricane reached the Downs, and at 7.30 P.M. the wind suddenly shifted to SSW., with increasing gale, so that by 1.30 A.M. of the 14th, the hurricane (as I had months before predicted for this day) swept over the metropolis, and with what violence is well known; while during the 14th, it desolated many a home on the east coast, and it was not till the evening of the 15th that its fury had abated at Shields and the northern ports, leaving a melancholy trace of its destructive power wherever it had been. Now this is either a truth or a fiction. If truth, no time was to be lost in examining the question of weather as traceable to lunar or other influences. In compliance, therefore, with the suggestions of underwriters and others, I issued this little book, in order to enable the public to benefit if possible by my labours of many years, and indeed to protect myself from all imputations of empiricism accordingly, by placing before the public the real state of the question of 'foretelling' weather, in its altered and improved state.

In order to show the *tone* and *manner* in which I first troubled the public with this my question, which has become European, I reproduce the following letter from me which appeared in the 'Nautical' for January 1860. I sent with it, for the private satisfaction of the distinguished editor,* details in proof of accuracy:—

'Sir,—Will you allow me to make known through the "Nautical" (as the readiest mode of communicating with those whom it most concerns) that I have for some time observed facts which appear to be incompatible with the received opinion that the moon has no influence on the weather.

'I have noticed during the past year (and reference to other years confirms my suspicions) that the moon never crosses the earth's equator without there being a simultaneous disturbance of the barometer or thermometer, or both.

'This assertion is, of course, open to challenge. If my ground be proved untenable, I am ready to retire. But from what I have noticed, it will not be yielded without a struggle. The moon's undoubted action on the tides is not to all places simultaneous, because (irrespective of other influences) the varieties in the form of coast-line may retard or even facilitate the motion of the tidal wave. It is not, therefore, absurd to suppose or assume that in like manner certain influences, at present unknown, really exist, which modify the moon's action on our atmosphere. Nor need we search far into the question before a suspicion is produced as to such influences, when we consider the possible effect of moon's age, perigee, or apogee, the earth's aphelion or perihelion, &c., &c. These speculations are left for those who have more time for investigation.

'My object at present is to state, that for my own personal convenience, I have projected the barometric curve to scale for every three hours from 9h. A.M. till 9h. or 12h. P.M. daily during several years, and I have found that without projection the practical use of the barometer is very much limited, and especially in stormy or unsettled weather. Indeed, by watching habitually a frequently projected height of barometer, we find "prognostics" which a mere look or glance at *numbers* would fail to present.

* Captain A. B. Becher, R.N. F.R.A.S.

'I would propose the following question:—

'If I take any period of consecutive months or years, and on every occasion during that period notice the barometric curve as registered while the moon was crossing the equator;—if, moreover, against the various curves so registered I place the identical words then written as indicating weather and its changes;—if, for instance, I take a period of eighteen months while at Rock Ferry, in Cheshire, between January 1856, and June 1857;— and as a counter test, or "corroborative," I take the last twelve months at Sheerness, and in both these cases and under the same circumstances find the same results invariably apparent—am I justified or not in expressing publicly a desire to see so interesting a subject properly investigated?

'The results may in general terms be thus briefly described:— The moon seems never to cross the earth's equator without there occurring at the same time a palpable and unmistakable change in the weather.

'*Such changes most commonly are accompanied either by strong winds, gales, sudden frost, sudden thaw, sudden calms, or other certain interruptions of the weather, according to the season.*

'The most remarkable circumstance (next to that of high winds prevailing at the times of the moon's equinox) is that of the occurrence of most violent winds, which are apparently due to the moon's influence, *happening about two days after the day of the moon's crossing the equator.*

'If not encroaching too much on your space, I will give an illustration:—If any of your readers will take the trouble to project the height of the barometer for the 23rd October last, as observed at three or four times during the day, they will see a decided bend in the direction of the curve. By following onward this curve for two or three days, it will be found that about two days after (viz. on the 25th) began that terrible and disastrous gale which for years will be remembered by seamen as the "Royal Charter Gale." And it is remarkable, and in perfect accordance with my general assertion, that while the moon was actually passing the equator on the said 23rd, the weather changed from the "clear sky" and "fresh wind" of the day before, to a calm, with "dull" and very cold "drizzling rain," rapidly succeeded by snow and all the wars of the elements, which are described in the " Shipping Gazette " as " severe as any yet experienced in this country."

'Another case in illustration will suffice :—The moon crossed the equator on Friday, 16th December. The wind had been blowing very strong from the northward for three days, and it is a fact well worthy of notice that on the Thursday (15th) so little appearance was there in the sky of a change, or, rather, so strong were indications of a continuing and increasing regular northerly gale, that we never saw so large an assemblage of fishermen and river barges at anchor for shelter under the lee of Cockleshell Hard, in the Medway, expecting a north-westerly gale. Instead, however, of the gale so generally looked for by the oldest and most experienced sailors, no sooner did the moon near the equator, than, on the 15th at P.M., the wind rapidly moderated, and at the precise time of the lunar equinox (if one may use that term with the moon), a dead calm set in, the intensity of the cold increased, and the wind, which had been blowing from the NNW. for a few days, shifted round to SSE., where it (17th) now remains; while the barometric curve for the 16th evidences some disturbing cause in a manner too strongly to be doubted.

'S. M. SAXBY, R.N.'

In the following March number I thus continued (and *it will be seen how careful I was not to wound the feelings of those who differed from me*, and how cautiously, though confidently, I developed my views) :—

'SIR,—In your number for January last, to which I would respectfully refer your readers, I was kindly permitted to use the following paragraphs, &c. &c.

'I further requested general and special attention to the barometer on or about the 12th and 27th of January instant.

'Preferring a consideration of facts to speculative arguments, I will accordingly notice what has actually occurred since my last was put into your hands on the 19th of December : reminding your readers that I only offer remarks upon observations taken by myself at one place, viz. Sheerness.

'Reference to a diary of the weather and to a projected diagram of the barometer for December last, will show that for a period of about eight days previous to the 29th of December, the weather had been nearly one continued calm, interrupted only by a very moderate breeze on Monday 26th (which, by-the-bye, I believe to have been a hurricane somewhere), and a light pleasant breeze on the 28th; but it is at least remarkable as in connection with the above abstracts, that on the 29th, 30th, and 31st of December, the wind, which blew merely a smart gale at Sheerness, was so excessively violent in some parts of England, that in Wiltshire, at Calne for instance, it is supposed such a destructive storm had never previously visited that locality (see "Liverpool Mercury," 7th of January). Now the gale was at its height on the 30th—the very day on which the moon crossed the equator.

'Again, I noticed that for about eight days previous to the moon's equinox at midnight of the 12th of January, calms and uninterrupted light winds prevailed from the westward; but it would seem that no sooner did the moon approach the equator than the characteristic disturbance of the barometer occurred, although, from some cause as yet unknown, it was smaller in amount than on most occasions; but it was attended with a sudden sharp frost and distinct change of wind from SW. to NNE.; the wind again on the 14th resuming its southerly quarter. But that which so distinctly bears upon my assertion as quoted above, is the further remarkable circumstance, that shortly after the moon's equinox on the 12th, the barometer began to descend (and thus constituting a "change" to which I have referred) until midnight of the 14th (that is to say, about two days after the equinox), when the wind at 0h. 15m. A.M. of the 15th suddenly rose and rapidly increased to so strong a breeze that during the 15th the foul-weather flag was hoisted at the steam guard-ship in the Medway.

'The moon next crossed on the 27th of January, when again the calm weather of the previous three or four days was interrupted early in the morning of the 27th by a change of wind from southerly to N. by E., and the wind rose at Sheerness only to a fresh breeze, while the barometer curve on my diagram indicated a terrible gale not far distant—nor was this incorrect, for the dreadful NNE. gale on the north coast of England at this period was wrecking above seventy ships, and depriving of life some twenty-five of our fellow-men: the weather next day resuming its previous calm, and in further accordance with my previously quoted assertions, the thermometer fell very considerably.

'The next crossing happened on the 9th of February instant. The wind had been for a few days before this stormy from the westward. On the 8th it blew fresh from WNW.; but early in the morning of the 9th (the day of the lunar equinox), the wind suddenly shifted to the NNE., snow and hail fell before sunrise, and a very sharp frost set in. Two days afterwards the wind (for a few hours only, as if it were from some interruption) returned to the south.

'Sufficient having been said to illustrate my meaning as to Lunar Equinoctial Gales, it may be convenient to your readers if, for their ready comparison, I recapitulate, taking the last nine periods of the moon's crossing the equator.

'1859, October 23rd.—Change from fair weather to sleet and snow. The Royal Charter Gale set in two days afterwards.

'November 5th.—Very heavy gales, lightning, &c.—remarkable rise of barometer commenced two days afterwards, amounting in forty-eight hours to 1·18 inch.

'November 19th.—Change to easterly, very cold: decided barometrical disturbance.

'December 3rd.—Very great disturbance of the barometer. Change of wind from NNE. to SSW., with very heavy gale two days afterwards.

'December 16th.—Change of wind from NNW. to SSE.; very sharp frost. Only slight (but marked) disturbance in barometer.

'December 30th.—Terrific gales in different parts of the country (e.g. at Calne, &c.).

'1860, January 12th.—Change of wind to NNE., returning next day to south-westerly as before. Very strong wind two days after the equinox.

'January 27th.—Dreadful gale on the North coast, and change of wind from south to N. by E. and NNE.

'February 9th.—Strong gale and sudden shift of wind from

WNW. to NNE., with marked barometric disturbances two days afterwards.

'*I may venture therefore to mention as characteristics of the periods referred to—high winds—shifts of wind generally with colder weather—disturbances two days after the equinox, &c.*—and as I could go through a list of fifty such consecutive crossings and see the same result, *I trust I have not without some justification* presumed to combat opinions held by those whom the civilised world revere, and to whom I would respectfully commend the subject.

'If the moon influence our weather when crossing the earth's equator, we might likewise expect to find some corresponding disturbance, in a greater or less degree, when she is on what I may call by analogy the " stitial colures : "—such is absolutely the case.

'It must not be supposed that I would expect changes of a like nature to occur either at the same moment at different places, or that such changes should at the same place be always similar in character. It would be unfair in considering this important subject to reject in our investigations the efficacy of local influences, such for example as the neighbourhood of high land—whether the place of observation be on the north or south side of high hills, especially when proximate to large bodies of water, &c.; but I beg to submit that the facts to which I have referred are within the power of the multitude to corroborate or refute.

'*If the circumstances to which I refer can be explained as unconnected with lunar influences, I shall not on conviction feel at all ashamed to confess in your pages my misconception. They are at least extraordinary coincidences.*

'I am &c.,
'S. M. SAXBY, R.N.'

CHAPTER II.

ITS RECEPTION—THE 'WEATHER BOOK'—DOVÉ'S 'LAW OF STORMS'—ASTRO-METEOROLOGISTS — OPINIONS OF EMINENT MEN — PUBLIC INTEREST IN THE QUESTION OF WEATHER — FACTS OR FALLACIES.

WE have lately seen the subject of weather elaborately handled by Admiral Fitz Roy in his certainly very interesting and elegantly illustrated 'Weather Book,' a work which few can read without receiving an impression that its author is a man of indefatigable perseverance, and of very considerable experience as a navigator and observer. This work also gives opinions upon weather, which have of course gone into every grade of society as orthodox, and as if *informing the community upon all which was worthy of notice that had been done in the question of foretelling weather* up to the year 1863.

In the year 1862 Professor Dové's work had appeared as what it will long be considered to be, the text-book of the 'Law of Storms.' Never, therefore, has the attention of the scientific world been more definitely called to the consideration of weather and weather 'prognostics' than at the present time. So much the more, then, does it become me to clear the subject of all uncertainties and wrong impressions, both as regards myself personally, and with respect to my known opinions.

Having shown in the last chapter that I had acted with all courtesy towards those who might have differed from me, and even, to avoid controversy, had spoken of what I *knew* to be a discovery, as *possibly coincidences*, I am the more free to examine the opinions of others, especially when further experience of three years' close observation has given me better ability so to do.

Differences of opinion between individuals in matters of science, and particularly in so entangled and difficult a branch of it as meteorology, do not *of necessity* lead to acerbities. There is an open and conciliating course which, while it sufficiently *probes* the question in order to elicit its truths, is calculated to rather soothe than otherwise those whose opposite views may have been arrived at with as much honesty of purpose as my own.

Holding as I have done for years a public appointment in which consistency of character and judgment are a *sine quâ non*, and a certain clearness of perception in me as a man of experience is indispensable to my efficiency, the question of my accuracy and consistency become involved whenever remarks against the lunar theory of weather-changes appear before the community. It is prudent, therefore, now to publish the result of my experience.

I would first plainly declare that I do not, nor did I ever, believe in *astro*-meteorology. The laws of gravitation apply, of course, to all space as a universal law, but it is not difficult to demonstrate that its effect on the planets as parts of the solar system is quite another thing in amount from their effect on our earth's atmosphere. Believers in *astro*-meteorology fall into the error of estimating the *bulk*, and not the distance and specific gravity of celestial bodies. It is obviously of importance, for instance, to consider whether a planet has the specific gravity of our earth or of cork! or whether the distance be one quarter of a million of miles from us, as the moon is, or, as in the case of Jupiter, nearly five hundred millions of miles! with, moreover, the fact that attraction decreases inversely as the cube of the distance!

I had the misfortune to differ from all the greatest men

of Europe in my opinions upon lunar influences on weather. In my idea a man who like myself wrote from *expressed* conviction deserved sympathy and courtesy; and indeed 'real philosophers' thought so too, for with great kindness I was advised by one whom all the world of science reveres, not to follow up what he considered to be a fruitless pursuit, as the influence of the moon on our atmosphere had been so often tested that various theories connected therewith had broken down when fairly investigated. I shall ever feel grateful to him for his honest reply; others maintained silence; while some, and among them one who is acknowledged to be one of the lights of meteorological science, urged and encouraged perseverance, saying, *that theory offered no obstacles to my views.* But from among all these truly great men not one sarcastic word reached me, notwithstanding my apparent presumption.

Admiral Fitz Roy also received me courteously. I had considered it due to him, as the head of the practical department, that I should offer my new discovery for his scrutiny, and accordingly in February 1861 called at his office, and while he refused to hear a description of my experiences in the supposed new theory, under the very natural plea that with all the appliances of the country at his command, he could not possibly need any help or information from others, lent me instruments to assist in my further investigations.

But the question of weather, like others of scientific importance to a great maritime nation, is open to all the world; it is not the custom of the age to yield to the unchallenged dicta of even the wisest and greatest. I claim indulgence, therefore, in offering the following.

Anxiety to make known what I believe to be of great public value has caused many thousands of total strangers to appeal to me on the subject of coming weather. The piles of letters received by me within the past two years on the subject of meteorology renders it proper that I should disavow all desire to interfere with the really difficult duties of the naval head of the department of practically considered meteorology. I have not only too much independence and too much self-respect to presume on his path, but too much respect for the Board of Trade as a valuable national institution to afford even the *appearance* of meddling with its officers.

Mere personal opinions on such an intricate subject as weather would be of doubtful value if unsupported by facts. But the difficulty is to know what are facts and what are fallacies. I will give an illustrative example:—If I hold my watch beyond the distance of a foot from my right ear, I cannot, let us suppose, hear it 'ticking,' while its sound is distinctly heard by my children at a greater distance. Shall I then, because of my age and experience, and because of the like qualifications of others, we will say as deaf as myself—shall I, in defiance of younger perceptions and their keener sense of hearing, declare that when so placed my watch stops, *because I cannot hear it?* It has been just so with the mental perceptions of many who could *see* no connection between the moon's influence and the weather, *while I could plainly trace it.* I am therefore placed in the onerous position of being obliged to prove all I say upon the subject, and the more especially as one of the most illustrious and liberal of philosophers has just again denounced all idea of lunar agency in atmospheric disturbance as fallacious (to which we shall further refer).

Let me first show that this adverse opinion generally exists. In a work, published in 1863, by a member[*] of the Meteorological Society, occurs the following:—

'Predictors of weather have existed in most countries; and means of foretelling what the alterations of the atmosphere would

[*] Mr. Thomas Hopkins (Longman & Co.).

be have always been, and are yet, sought with great avidity by man. Our satellite has generally been considered, and is still extensively believed, to have great influence over the weather. And considering the state in which meteorology has been left, it is not surprising that such a belief should continue. The moon is known to cause tides in the ocean, and therefore is likely to be believed to have influence on the atmosphere, until meteorology becomes a science, and proves that it has not. Notwithstanding, however, that great industry has been displayed to trace some connection between the moon and weather, none has been discovered.'

That such opinions are not singular, and prevail in other countries, is shown in a pamphlet published in 1863, from Professor Buys-Ballot of the Royal Netherlands Meteorological Institute, and translated into English by Dr. Adriani, who says:—

'For though I feel quite convinced that it is possible that at some future period the state of the weather will be determined in advance with the same accuracy as we now predict the correct moment of an eclipse of the sun or moon, we are bound to say, that as regards the weather, *we are not at present so advanced yet*.'

And adds:—

'Whatever the opinion may be on the science of meteorology and its progress (and we should bear in mind that many other important truths have had to pass through *a period of mockery*), this,' &c., &c.

Mere difference of opinion excites a healthy investigation, but there is a course, which, in too many instances, deadens the ardour of the honest observer, and protracts till after years the difficulties which beset many an important inquiry:—thus truth suffers, and, singularly, that very truth, moreover, which those who thoughtlessly utter the discouragement, would in other respects almost sacrifice a limb rather than *deliberately* stifle.

CHAPTER III:

THE WORLD'S DISBELIEF IN THE MOON'S INFLUENCE ON WEATHER AS SHOWN BY THE 'WEATHER BOOK.'—'LUNI-SOLAR' THEORY.

AND first as to the world's *disbelief* in Lunar influence on weather.

The 'Weather Book' opens with the following declaration, at page 3, that—

To 'solar heat acting diurnally on the atmosphere as our globe rotates, combined with its absence or COLD,* and with unfailing gravitation, *it is sufficient to ascribe all our atmospheric conditions* and changes *without* AT PRESENT *drawing in any powerful* LUNAR *influence, or other planetary relations.*'

On page 4 it is acknowledged, however,—

' That the moon, as well as, and probably much more, than the sun, causes a tidal effect in air, due to gravitation, *cannot be doubted.*'

But that—

' Recurring daily causes immediately referable to solar heating *so greatly overbear this as for it to be almost undistinguishable.*'

On the same page (4) it says also :—

' It is remarkable that ' Astro-meteorologists' and ' Lunarists' have not observed that their supposed causes of weather must, if existent, affect entire zones of our atmosphere, in diurnal rotation, instead of one locality ALONE, and that *such results are not proved by the facts observed.*'

* The italics are mine in all these quotations; the larger type is in italics in the original.

And again on the same page* (4) it says—

'When persons who attribute changes of weather to the moon are asked,' &c., &c.: 'therefore no satisfactory information can thus be gained, *and we remain baffled.*'

The poor 'Lunarist' is then left at peace until he approaches the 213th page—meanwhile reading in the Weather Book much that is *highly interesting and useful;* but after it having referred, at page 212, to extensive action in our atmosphere, at 213 he is brought up all standing by the following:—

'In such grand disturbances as these, the *Lunarist and the Astro-meteorologist* should endeavour to trace influences of moon and planets. Welcome, indeed, would each proved effect of either be—duly eliminated from masking effects of other causations.'

While wishing to honestly guard myself from any misapprehension of the author's meaning, I submit that the above proves the naval head of the Board of Trade to have been, when he wrote it, no advocate for 'Lunar influences' on weather, we may safely say that the gallant Admiral was no 'Lunarist.' Indeed, to have ranked him below 'alchymists and astrologers of old,' would have been impertinence—*insult.* I am, therefore, correct in declaring that the scientific world (taking Admiral Fitz Roy as its mouth-piece in meteorology) *disbelieve* in Lunar theories of weather.

Turn we now to the next consideration—the next curious, anomalous, and antithetical proposition, viz. that the author *does likewise believe* in 'Lunar influences' as affecting weather.

After saying at his page 4, that—

'All the effects *caused by gravitation* towards sun or moon have been found, by repeated observation, to be *so greatly overborne*

---

* Pages referred to are those of the first edition.

or masked by recurring daily causes, immediately referable to solar heating or electrical action, *as to be almost undistinguishable even at places supposed to be most eligible for observation.*'

He says at page 244 (and in effect elsewhere)—

'The moon also acts on every particle of air, as on water and earth, by universal *gravitation.* Tidal effects must therefore be caused *by the moon* and by the sun in earth's atmosphere, *and their scales may be* LARGE (it is submitted) in proportion to its depth and extreme mobility.'

Now, with such apparently unsettled opinions in the same author, at what conclusion can we arrive ? except that while writing page 4 he was *not* a Lunarist; while at writing page 244 he *was one!* In other words, the truth had dawned upon the gallant Admiral in the interim, and with the frankness of a sailor he confessed it—*but in his enthusiasm forgot to state to whom he was indebted for the said truth.* But let me be cautious how we attribute great and sudden change of opinion to one of so much and such long experience. I should be ashamed to distort any phrase or sentence to suit my own interest, nor would I do so, but I am obliged, unwillingly, to quote other passages from the so widely-known, and in so many respects, excellent 'Weather Book.' It is only to be regretted that absolute necessity exists for these quotations. Singularly, in the very next page (245) the author quotes the Newtonian theory (on which *precisely* mine is built), as what we ought to find proved by observation, but declares quantities and effects thus found by experiment to be *almost insensible!* When he wrote page 245, therefore, he could not have been a 'Lunarist.' But, more wonderful, in again the very next page (246), he, in reasoning on the moon's action, becomes again a 'Lunarist.' I may also add *enthusiastically so.* I wished to avoid long abstracts, but really the reader cannot be expected to yield credence to such assertions without *what I strongly recommend,* an actual perusal of the 'Weather Book' itself. But, as I shall have, further onward, to call attention to the passage, it will be better once for all to give it from page 245.

'But their greatest effects should be not only thus *in accordance with the Newtonian theory* (supposing earth entirely smooth and covered by an ocean), but all the lunar and solar periodicities, of apogee, perigee, and declination, would have proportion to effects, and the extremes of all actions combined together, *would be when sun and moon, in perigee and in syzygy, are in or near the equator.*

'Next in importance to these would be the actions, extratropically, of both luminaries *in extreme similar declination, and likewise in perigee.* Now what are the facts as hitherto observed?

'Such observations as have been made, barometrically, would appear to have almost demonstrated the *absence* of any direct solar tide that causes more than a few inches (less than a foot) of vertical effect, and of any *lunar* tide that occasions a rise of more than a few feet of air, *quantities seemingly almost insensible,* and quite inoperative, one might consider, as principal agents, *much less as prime motors of atmospheric currents.*'

And after saying at page 246 that *statical* measures cannot alone elicit 'all the facts of this important case,' &c.; and after considering the 'moon's' action, *alone,* and supposing her in the Equator, he adds:—

'RECURRING PERIODS of about fourteen days (semilunar), of seven, and of three or four days, have been traced, *however masked* and irregular, *more or less synchronous with the moon's phases,* OCCASIONALLY, and then, for a few turns, rather correspondent, therefore *evidencing some kind of connection*; but a VERA CAUSA seemed to be wanting for an explanation. One might say, indeed, after entering a little further into the apparent consequences and connected relations of this ENTIRELY NEW, AND EVEN TO THE WRITER (Admiral Fitz Roy) STILL MOST STARTLING VIEW (SO SATISFACTORILY DOES IT SEEM TO ELUCIDATE SOME OF THE GREATEST DIFFICULTIES OF METEOROLOGY—SE NON È VERO, È BEN TROVATO' (fairly translated thus: *truth or untruth, I like the notion*).

Now, if I 'startled' the worthy admiral with my theory, he certainly has astonished me in return; for, at a few pages onward, he seems, in somewhat vacillating language, to attribute the

'Popular belief in a connection between weather and the moon, to " *casual coincidences* "' (page 252) ' *rather than to " scientific facts.*"'

While it is most extraordinary that in the very next page (253) he gives an opinion upon what should be the consequences of the 'moon's greatest tidal action' in the atmosphere, and declares that —

' *The facts observed, to whatever cause attributed, do correspond exactly to these postulates.*'

And still more strongly to confirm singular changes of opinion, he in his same page 253 says :—

'What causes such equinoctial disturbances? Not the mere fact of the sun's astronomical position? No; *the united tidal action of moon and sun upon* THE WHOLE ATMOSPHERE, *which then is a maximum force. Lateral offsets,* streams overflowing towards each pole, and, as they go, preserving more or less momentum, are at those times more powerful, and their effects are more FELT EVERYWHERE.'

It is worthy of note that in this—the said author's declaration (which is in accord with the Newtonian theory), viz. that the

'United tidal action of moon and sun upon the *whole atmosphere*' (the italics here are his own) ' produces effects *which are more felt everywhere,*'

is in direct contradiction to his assertion at page 245—that such agencies produce effects which are seemingly *almost insensible.* Such seeming contradictions apart, it is a consolation that the gallant Admiral has actually been 'caught' (see his p. 253) by the apparent truths contained in a mass of evidence which had *caught me previously,* the present difference between us being that I am still held fast by them, while he has wriggled himself clear of their unorthodox entanglements, and *means to have nothing to do with them.* If therefore any obloquy can possibly attach to me as a 'Lunarist,' the gallant Admiral is not exempt from participation, although his present opinion is that of the multitude.

Now, as an Englishman, and a loyal subject of our beloved Sovereign—the character of every distinguished man in her noble service, to which I have the honour also to belong, is dear to me; *but can I by longer silence be expected to yield my pretension to a discovery which has cost me through years much trouble and expense?*

The Admiral is not singular is disbelieving in the moon's action upon the atmosphere. I say '*disbelieving*,' but from what immediately follows, the reader would be inclined to think me incorrect in saying so, had he not so recently quoted Sir John Herschel and others in declaring that *enough is not yet known* to warrant a reliable prediction of weather—for many days in advance.

The following is another justification of my no longer submitting to be set aside in the public question of Lunar influence. 'Chambers's Journal' of July last (six months after the publication of the Weather Book), quotes at page 16 the following from Admiral Fitz Roy's report—his official report, just then issued—'Some insight has been obtained of the manner in which atmospheric changes are occasioned, and air currents or winds *set in motion* around our world, *chiefly by the sun and moon.*' But the writer of the 'Chambers's Journal' adds the remark; 'In this particular, however, he (the Admiral) is careful to state that the highest scientific judges do not yet accept the (qy. his?) Luni-solar theory.'

The public must not be misled by the phrase 'Luni-solar,' and imagine that Admiral Fitz Roy means thereby that he has discovered anything differing by the *slightest shade* from what I call a 'Lunar' theory. *The two are precisely the same.* What he calls Luni-solar, is the joint or combined action of sun and moon at certain periods of the *moon's* orbit (hence my term 'Lunar'). Take for example what I attribute to the new moon—as of course a true Luni-solar agency. The substitution of the term is calculated (unintentionally) to mislead the public, which, of course, the admiral would be sorry to do. But how can he accept a Luni-solar theory? when he, in his 'Weather Book,' at page 245, says :—

A *lunar tide produces an effect* ' *in quantities seemingly almost insensible* and *quite inoperative,* one might consider *as principal agents,* MUCH LESS AS PRIME MOTORS *of atmospheric currents* ! ! '

I will, in a new chapter, prove that the author of the 'Weather Book' is not singular in his *disbelief* in the moon's *appreciable* action in producing changes of weather, but there can be no doubt that when he wrote page 4 he was *indisputably not* a Lunarist.

and page 213 he was indisputably *not* a Lunarist.
,, ,, 244 he was *certainly* a Lunarist.
,, ,, 245 he was again *not* a Lunarist.
,, ,, 248 he was an *ultra* Lunarist.
,, ,, 253 he was still a Lunarist.

But recent letters in the *Times* show that he is not now a Lunarist.

CHAPTER IV.

SIR JOHN HERSCHEL'S OPINION IS THAT OF ALL PHILOSOPHERS — MY CLAIM TO THE DISCOVERY OF THE LUNAR THEORY OF WEATHER ANTICIPATIONS.

THE illustrious individual, whose opinion I next quote, is, in his world-wide fame so justly appreciated, that it seems like sacrilege to doubt his accuracy upon *all* points of practical science, while to arraign his conclusions as to 'lunar theory' would almost excite the pity of the reader towards me for my apparently insane presumption; and yet we must not sacrifice to any altar but *truth*. The question under consideration is a momentous one; and I, with the greatest unwillingness, refer to an opinion just published by Sir John Herschel in the serial called 'Good Words.' In the present (January) number, under the head of 'The Weather and the Weather Prophets,' to which his respected name is attached, occur the following. At page 58, he says: —

'It is to be borne in mind, however, most carefully, that all such indications (from barometer, thermometer, &c.) are to be received as valid (*pro tanto*) *only for a very brief interval in advance*; *and that the "weather-prophet"* who ventures his predictions on the great scale *is altogether to be distrusted—a lucky hit may be made*: nay, some rude approach to the perception of a "Cycle of Seasons" may possibly be attainable.

'The moon is often appealed to as a great indicator of the weather, and especially its changes as taken in conjunction with some existing state of wind or sky. As an attracting body, causing an 'aërial tide,' it has of course *an* effect; but one *utterly insignificant as a meteorological cause,* and the only effect distinctly connected with its position with regard to the sun, which can be reckoned upon with any degree of certainty, is its tendency to clear the sky of cloud, &c. &c.'

Now the above accords (and I must in fairness mention it) with the statement of Admiral Fitz Roy at his pages 4 and 245 — *a pretty position in which to find myself in the eyes of all 'Europe'!*

But there are two sides to the question. In my various writings accuracy has been my aim upon all occasions, and even in this I hope to show that facts are in my favour. *If so, what a triumph!*

But it is not because an author disbelieves or believes in the correctness of my conclusions as to Lunar influences that I quote him, for it is an open question. The object of my noticing the above portions of his work was to show the necessity of my now asking — *Who,* IF NOT MYSELF, startled him with *an 'entirely new' idea, the very one to which I had been so long carefully giving wide publicity?* In the 'Nautical Magazine' (as already referred to), in the 'Standard,' the 'Sheerness Guardian,'—by private handbills and special notices sent, at my own expense, round the whole coast of Great Britain as particular warnings—by public announcements and by private letters in reply to total strangers, many of them the most experienced of observers, naval officers and merchants, navigators, agriculturists, clergymen in numbers (I have piles of their letters), until every nook and corner of the kingdom contained my weather list—had I issued warnings as to coming weather; and, indeed, in June last, as a matter of courtesy, I sent a copy of my list of predictions for the present winter to Admiral Fitz Roy himself.

I am, therefore, in a position in which to ask thus publicly why I was deprived of what in the author of the 'Weather Book' would have been a graceful, and consistent, and due official recognition of my claim to the discovery of a Lunar system which was in his estimation, as I have proved, so worthy of note from its having borne the test of observation, 'and as seeming,' at the time he wrote his page 246, 'to elucidate *some of the greatest difficulties in meteorology*'?

I had, moreover, in February 1861, and I must repeat it, a personal interview with the gallant Admiral, at which I felt it to be my duty to place at his service (as the official head of practical meteorology) whatever results my labours through years had produced, and they were freely and 'loyally' offered; but his expressed conviction that, *with all his immense official resources, he could not need any assistance*, was a palpable rejection. (N.B. I left his office with even my diagrams unopened! It is but just, however, to state that it was on that occasion he lent me the standard barometer, &c. which I have found so useful.) But, further, he seems (unwittingly, of course,) to have regularly adopted my lunar theory, in the following words:—

'They' (such effects) '*vary with the declination and distance of the moon.*'

This is the very pith of my theory; and he immediately adds:—

'During the moon's passage in her orbit from quadratures to syzygy, her action on air currents should increase, and conversely, when she has great north declination it ought to be greater here than when she is far south, *and when in perigee greater than in apogee.* Tabular records show that such are the facts.'

But the clenching of the whole matter is in his official declaration that '*Tabular records show that such are the facts.*' Thus leaving it open to any future unscrupulous or careless reader of his Book *to quote him as the first* who noticed such tabulated facts.

Therefore, **ADMIRAL FITZ ROY, WHEN HE WROTE THAT, MUST HAVE BELIEVED FIRMLY IN MY LUNAR THEORY AS PROVED TO BE ACCURATE.** But I again ask *who* proved it? *If he did, how is it that Sir John Herschel twelve months afterwards denies appreciable Lunar influences altogether?* The fact is, neither has yet heard the ticking of the watch' — *but I have.* The listening of the universally appreciated and illustrious veteran philosopher has not detected it, while comparatively a mere youth in science has chanced to do so distinctly.

I think I can clear the whole question of its present difficulties, and my hope is that in the end *all* will be found right, according to the means each has had for the basement of his opinion. Neither will have to regret a little mistake or difference in such an intricate subject if shortly we can send out weather-lists to sea to last through a whole naval commission, or if we can so warn the coasts and shipping *at sea* as to save *more* lives and *more* property.

No one is more able to carry out and improve upon a theory once established than Admiral Fitz Roy—no one more energetic and indefatigable, as I shall yet, in Chapter VI. have real pleasure in showing. We shall yet see both him and the distinguished philosopher whom I have quoted, acknowledged 'Lunarists.' I will not say that the *storms*, but the periods at which such are *most likely to occur*, will be indicated years in advance, and upon principles first given to the world by the immortal Newton. So great will be the joy of the multitude, that promenaders will be humming and whistling *in security* their 'Mackintosh Galops,' 'Umbrella Quadrilles,' or, I should not wonder, their 'Cyclone Waltzes.' But I am anticipating. The reader possibly will have already smiled in incredulity at my supposed presumption. My convictions, however, are *strong*; on what foundations they rest will appear in succeeding chapters.

CHAPTER V.

CAUSE OF POPULAR ERROR EXPLAINED — THE ANCIENTS AS WEATHER-WISE AS OURSELVES — PROOFS — PROGRESS (?) DURING 1800 YEARS.

I WILL now turn to the subject of such proofs as seem to warrant all my hopes of successful vindication of my 'Weather System.'

But first, let it be granted as demonstrable from the preceding pages, that while I write the scientific world *strongly* repudiate the notion that lunar influence can affect weather. If I can show indisputable evidences that the moon does disturb our atmosphere, and therefore our weather, I only ask for my opinions to be fairly recognised.

In attempting to account for the singular bias in the minds of not merely Sir John Herschel, of whom one can but speak with reverence, but of all philosophers, against the moon's agency, *I can only attribute it to a defective mode of using the barometric indications*; for, although it has been almost universally acknowledged that at times that instrument is faulty in its warnings, *we have registered and tabulated as if it were infallible*, as if it were sensitive to *every* kind of atmospheric disturbance!

In ignorance of one cause of atmospheric disturbance, viz. change in electric tension of the air as *influenced by the moon's position in her orbit*, we have hitherto been without the means of accounting for such vagaries in the barometer as are apparent in all well kept meteorological diaries. We have seen bad weather occur when the barometer gave no previous warning whatever. The oldest sailors will often have been at a loss to account for occasional failures also in the usual sky prognostics. Yet (as I have said) we tabulate, and compare, and deduce, as if the barometric height were the sole means of interpreting the state of those influences which determine coming weather.

Now, my early attempts in meteorology were to search for *periods of change*, and during a large portion of a lifetime of observation I had often regretted the seeming impossibilities of getting a clue to success.

We live in a vastly enlightened age as regards practical science, but I ask *where in the scale of comparative development stands meteorology?* Beyond what we owe to Admiral Fitz Roy for the introduction of telegraphy, *what more do we know of coming weather than (for example) we did when I was a young man of twenty?* I say deliberately and advisedly that from my early intercourse with the nautical world, and with its men of large experience, I knew as much of sky prognostics and the supposed indication of weather changes from physical evidences when I was young as I do now! We have more accurate instruments, and an infinitely larger number of observers, but I believe that our grandfathers, notwithstanding, were as weather-wise as regards forecasting weather *without the aid* of instruments (except the barometer), as *we are with* all the appliances of those beautiful self-registering facilities which so imposingly furnish our observatories! I mean to say that, with a glance at the old barometer, and a look around the horizon at sunset or sunrise, &c., they could quite as well guess the kind of weather to be looked for next day as we can do now from the study of instruments *added to* the study of such ordinary *appearances* as every experienced man will attend to. *I have noted materials for the proof of this assertion,* and only suppress them from obvious considerations, ready, however, for production if required. I assert that we cannot authoritatively, and as a rule, predict *from instruments* with certainty for twenty-four hours, or even less. I say it not only from my own opinion, but moreover on the authority of Sir John Herschel and Admiral Fitz Roy. Indeed, to use the words of the great astronomer published within a few weeks of the date on which I write, ' meteorology, so far as prediction is concerned, may be regarded as a science still in its infancy.'

Some of my readers will be surprised to hear how much of what we call ' prognostics ' were known so long back as in the days of the ancients. Whatever Professor Dové may have simplified, and Admiral Fitz Roy may have illustrated, prognostics in present use chiefly owe their origin to remote antiquity. Signs from vapours, from clouds,* from the dew, from the face of the sky, from the appearance of the sun, moon, or stars, from winds, from nocturnal meteors, from even the animal creation, from flowers, &c., we still use these by the side of Dové's able remarks on air currents, and ' equatorial ' and ' polar currents,' with much combination of theoretical considerations *which rather explain the nature of certain atmospheric phenomena than give us skill in predicting their changes.* The wary philosopher will still bid us treasure every fact connected with meteorology, and watch, and observe, and register, and compare, and tabulate, until the grand result shall be attained in the so long courted ' weather wisdom.'

But to return: In the first Georgic of Virgil, written about the time of the birth of our Saviour, we find the groundwork of all we still publish and repeat as valuable in our lists of ' prognostics.' Even as regards the signs from the animal creation, we are forestalled by that

---

* See the 'Barometer Manual.'

author, as shown in the following beautiful lines by his translator Dryden:—

> '*Wet* weather seldom hurts the most unwise,
> So plain the signs, such prophets are the skies:
> The wary *crane* foresees it first, and sails
> Above the storm, and leaves the hollow vales;
> The *cow* looks up, and from afar can find
> The change of heav'n, and snuffs it in the wind;
> The *swallow* skims the river's wat'ry face,
> The *frogs* renew the croaks of their loquacious race;
> The careful *ant* her secret cell forsakes,
> And draws her eggs along the narrow tracks;
> Huge flocks of rising *rooks* forsake their food,
> And crying seek the shelter of the wood.
> Besides, the several sorts of *water-fowls*
> That swim the seas or haunt the standing pools,
> Then lave their backs with sprinkling dews in vain,
> And stem the stream to meet the promised rain.
> Then after showers 'tis easy to descry
> Returning suns and a *serener* sky.'

And again:—

> 'And *owls* that mark the setting sun, declare
> A starlight evening and a morning *fair*.'

And further:—

> 'Then thrice the *ravens* rend the liquid air,
> And croaking notes proclaim the *settled fair*;
> Then round their airy palaces they fly
> To greet the sun, and, seized with secret joy,
> When storms are overblown, with food repair
> To their forsaken nests and callow care.'

Indeed, the crow was much observed by the ancients as forewarning for rain. Virgil says, moreover—

> 'The *crow* with clam'rous cries the shower demands,
> And single stalks along the desert sands.'

It is better to show the *extent* to which the ancients had considered prognostics: it will appear that Pliny also (xviii. ch. 35) refers to the crow:—

'Et cum terrestres volucres contra aquas clangores fundentes sese, sed maximè *cornix*.'

(When land-birds, and *especially crows*, wash themselves, and are noisy near pools of water, it is a sign of rain.)

Even Horace adds his remark (Od. iii. 17):—

' . . . Aquæ nisi fallit augur,
    Annosa cornix.'

(Unless the aged crow, foreboder of wet, deceives me.)

And again, in his 27th ode, he refers to the *raven* as presaging rain, in these words:—

' Antequam stantes repetat paludes
  Imbrium divina avis imminentum,
  Oscinem *corvum* prece suscitabo
    Solis ab ortu.'

(I will by prayer, from sunrise, arouse the croaking *raven* before the bird that foretells approaching rain revisits the standing pools.)

And again, as regards prognostics from the skies, still in present use, which we obtained from Virgil, we find in various parts the following:

First, as to the moon:—

' Observe the daily circle of the *sun*,
  And the short year of each revolving *moon*:
  By them thou shalt foresee the following day;
  Nor shall a starry night thy hopes betray.
  When first the *moon* appears, if then she shrouds
  Her silver crescent, tipp'd with sable clouds,
  Conclude she bodes a tempest on the main,
  And brews for fields impetuous floods of *rain*;
  Or if her face with fiery flushings glow,
  Expect the rattling *winds* aloft to blow.

> But four nights old (for that's the surest sign),
> With sharpened horns, if glorious then she shine,
> Next day, not only that but all the moon,
> 'Till her revolving race be wholly run,
> Are void of tempests both by sea and land,' &c.

Then as to the rising sun:—

> 'Above the rest, the *sun, who never lies,*
> Foretells the change of weather in the skies;
> For if he rise unwilling to his race,
> Clouds on his brow and spots upon his face,
> Or if through mists he shoots his sullen beams,
> Frugal of light, in loose and straggling streams,
> *Suspect a drizzling* day, with southern rain, &c.
> But if with purple rays he brings the light,
> And a pure heav'n resigns to quiet night,
> No rising winds or falling storms are nigh.'

As to the setting sun, he says:—

> 'But more than all the *setting sun* survey
> When down the steep of heav'n he drives the day;
> For oft we find him finishing his race
> With various colours erring on his face.
> If *fiery red* his glowing globe descends,
> High winds and furious tempests he portends;
> But if his cheeks are swoln with livid *blue,*
> He bodes wet weather by his wat'ry hue:
> If dusky spots are varied on his brow,
> And, *streaked with red, a troubled colour* show,
> That sullen mixture shall at once declare
> Winds, rain, and storms, and elemental war.'

His signs or warnings for a tempest are as follows:—

> 'For ere the rising winds begin to roar,
> The working sea advances to the shore;
> Soft whispers run along the leafy woods,
> And mountains whistle to the murm'ring floods;

And chaff with eddying winds is toss'd around,
And dancing leaves are lifted from the ground,
And floating feathers on the water play,' &c.

As to signs from trees:—

' Mark well the flow'ring almonds in the wood:
If od'rous blooms the bearing branches load,
The glebe will answer to the sylvan reign,
Great heats will follow, and large crops of grain;
But if a wood of leaves o'ershade the tree,
Such and so barren will the harvest be;
In vain the hind shall vex the threshing floor,
For empty chaff and straw will be thy store.'

Volumes might be filled with the various rules for weather which our forefathers possessed, and *from which we still select* for the instruction of the unlearned or inexperienced. So much, then, for the supposed *advance* of human knowledge in weather prognostics, or in the power of *forecasting* from them, during a period of eighteen hundred years, and up to January 1st, 1864!—what if my humble labours shall have introduced a change! We shall see.

CHAPTER VI.

THE BAROMETER AND PROFESSOR BUYS-BALLOT — DEFENCE OF ADMIRAL FITZ ROY'S OFFICE AND DEPARTMENT — APPEAL IN BEHALF OF IT — ANEROIDS AS REWARDS FOR SAVING LIFE, ETC. — DEFENCE OF TELEGRAPH WARNINGS — WEATHER WARNINGS — SHIPWRECKED MARINERS' SOCIETY AND NATIONAL LIFE-BOAT INSTITUTE.

WITH regard to the barometer and its shortcomings as a means of warning reliably, it is almost painful to read of the elaborate care bestowed by some distinguished meteorologists upon it. If, for instance, we take the work of Dr. Buys-Ballot, the Professor of the Royal Netherlands Meteorological Institute, we shall find a system proposed by him for indicating weather, which is founded *solely* on barometric '*deviations*,' and even declared to be 'preferable to the English system,' the former being said to surpass the latter in 'notice being duly taken of the contemporaneous and simultaneous difference of the readings of the barometer of at least five well-arranged meteorological stations among each other.' Now, not at all wishing to detract from the meritorious labours of the learned Dutch professor, and yet called upon to substantiate my views upon the subject of weather warnings, I can but fear that his system is built upon a quicksand, because we know that the mercurial column is totally insensible to many great atmospheric disturbances which result in gales and bad weather, and which from their peculiar nature do not affect the barometer. But let it be only once admitted that there are certain undetected causes which lead to storms, *without disturbing the barometer*, and a useful system of warning from that instrument alone is, I submit, not possible.

If persons doubt that such storms do occur, let them ask any meteorologist, or, which is better, let them consult the 'Weather Book,' or the 'Barometer Manual.' Indeed, the latter says, 'The barometer falls, *but not always*, on the approach of thunder and lightning, *or when the atmosphere is highly charged with electricity.*' Now, considering that electric changes are the *immediate* cause of all our gales, except cyclonic storms, which travel towards us from a distance (see onward), and that it consequently is a recognised principle in meteorology that the barometer cannot *at all times* be relied on, I say that *any system founded on barometric variations alone, must prove a mistake and useless.* I cannot, therefore, see the superiority of the Dutch Professor Buys-Ballot's system over Admiral Fitz Roy's and the English. As Sir John Herschel suggests, 'lucky hits' may occur under all 'systems,'—I would not insinuate that Dr. Buys-Ballot's predictions on the 21st and 22nd November 1862 (by-the-bye, the only case he refers to in his pamphlet) were accidentally successful, but the description of his method is remarkable. His translator, Dr. Adriani, says at page 19 :—

'The reading of the barometer was just then (November 21st and 22nd, 1862) rather high, viz. 771 millimeters: this, notwithstanding the meteorological people, predicted even on the 20th a gale from the westward, and, as is well known, that gale took place and was rather violent.'

Now, 771 millimeters correspond to our 30·355 inches. Had the Professor's system of barometric deviations only been adopted in England, it would have distinctly misled the public; for on referring to my diagrams I find that the notes of Commander Field, R.N. (to whom I am indebted for much valuable registration on the west coast of Ireland (Dingle Bay), shows that at the same time (viz. from 12 to 6 P.M. of the 21st) the heights of the barometer at both stations, Dingle Bay and Sheerness, were identical, viz. at each place it was 30·19 inches, therefore, as an illustration, no '*deviations*' between the two places could have forewarned the public as to any coming storm occurring at either locality. Now what did *my* lunar system say as to that period? Reference shows that it was one against which I specially *warned the whole coast, not only months beforehand*, but in addition, I sent particular warnings, about ten days beforehand, through the 'Sheerness Guardian' newspaper; for the new moon in perigee occurring at the time of the moon's attaining her extreme southern declination, justified expectation of a troubled atmosphere, as the result proved.

But, further, we should enquire as to the nature of the *deviations* which are said above to have indicated a gale on the 20th. At page 18 of Dr. Adriani's translation we find the following information:—

'Suppose the average or normal reading for a certain day to be 761 millimeters, while the barometer on that very day reads 770 or 752 millimeters; then, in the first instance the reading, as compared with the average, is 9 millimeters too high, in the other 9 millimeters too low; in other words, the deviation from the normal is +9 millimeters, or −9 millimeters. Nobody can fail to acknowledge that it gives a clearer idea of the indications of the barometer when one is enabled to say its deviation is +9 millimeters, instead of merely announcing that the barometer stands at 770 millimeters, for the former conveys an idea of what kind of disturbance has taken place in the atmospheres,' &c.

Compare this with what we find at page 24, where he says:—

'I wish you to observe that the reading of the barometer of the different stations in the Netherlands has, by careful observations, as well for height above the sea-level as for the differences in latitude of the one station from the other, been correctly determined to within $\frac{1}{18}$th part of a millimeter; but in England (Great Britain) *such exactitude does not appear to have* been reached,' &c.

Now in the Professor's table, we see that half that amount of deviation is sufficient to warn of 'danger' (but it equals only ·221 of an inch), and also that a deviation of 1·4 millimeter (equal to ·065 of an inch) justifies the warning of 'caution.' Surely this is insufficient.

Were Admiral Fitz Roy to adopt the Dutch Professor's system, he would need a pretty large staff of assistants, while he could gain little or nothing by it. To observe to the three-thousandths of an inch would indeed be close work even to correct for temperature and reduction to sea level, &c.

This reminds me of a duty that every meteorologist and every sailor writing at this period—nay, every man engaged in British commerce — owes to the gallant Admiral and his department. He recently in 'The Times' notified his need of an increased establishment. I say with sincerity—*By all means give it him.*

The terrible gales of the present winter will have engaged the thoughts of the humane on the subject. The hundred Yarmouth fishing smacks, &c., having been *caught* in the fearful hurricane of December 3rd last, nearly forty of which were lost, shows that with all the energy of the Meteorological Department of the Board of Trade, we are far from having accomplished an adequate system of weather warning. It may, in a great measure, be that sailors, and especially fishermen, are not easily moved to the abandonment of old notions.

If I remember rightly, Admiral Fitz Roy warned the coast (on December 1st) to prepare for a gale. I, too, on that day personally cautioned certain fishermen, as their smacks passed close under the stern of the 'Devonshire,' that it was 'going to blow *hard.*' I feared this from the *falling of the barometer, and all appearances occurring so near to my lunar period of the* 4*th, of which I had warned six months previously.* Any intelligent person possessing a barometer might have then read, in the rapid descent of the mercury alone, enough to sufficiently warn of threatening danger, for the barometer *never falls without a threat* more or less imminent; and yet what frightful devastation was caused along the coast!

I have used the aneroid for years, and it is a most valuable and convenient instrument for small craft where a barometer could not be fixed; indeed it may, I believe, supplant the 'barometer' as a more compact and, for practical purposes, a more handy and equally exact instrument. It is a pity the expense of manufacture cannot be reduced so as to bring the aneroid within the means of poor fishermen. The distribution of a hundred or two around the coast among men who have been steady, and especially *to smacks and boats which have been successful in saving life* and property, would gradually introduce them, and the cost to the donors would be a trifle compared with the advantages which must result. It needs scarcely more care than a compass, and is not easily put out of order, or is very easily repaired by the maker. Oh that I could spare two or three hundred pounds for such a benevolence.— What a happiness many a one has it in his power to secure to himself and others, at small cost!

Those who at present so imprudently ignore a telegram would then be taught to believe first in the instrument and then in both.

Indifference to warning is unaccountable. Let us, as a parallel case, suppose men were working upon a railway line, and a watchman were posted (as is often done) on an eminence at some distance, whose duty it was to warn of an approaching train. What should we say if on a *danger signal* being given by him the men refused to leave their work, and were consequently scattered in mutilations and death ? And yet, strange to say, coast-men are continually guilty of like imprudence. If Admiral Fitz Roy says by his drums and cones, 'Look out! *I can see* (by my barometer) an approaching storm; it has already reached the western coasts and *is advancing*,' it is surely culpable to put to sea or to neglect shelter *unless well prepared for extremities*. Indeed, I do not see how the Admiral's telegram cautions *can* be otherwise than correct generally.

At present, however, these official warnings apply only to the coasts; those who may be beyond an hour or two's sail from land are beyond the reach of them. What if I can supply the warning for those so circumstanced! (I am about to do so.) But we must not be misled by supposing that the barometer alone is to be relied on. It is a general opinion that no *great* hurricane occurs without being predicted by the fall of mercury in the barometer. So far as regards heavy southerly gales this is about true, but does not always apply to heavy northerly gales, nor to those which I am about to show to be *immediately* the results of electricity. Now, if we ascertain the *periods* at which such may be confidently expected, we need little more in order to entitle us to the reputation of sound 'weather wisdom.' Our present weather wisdom must be defective, for no one can contemplate the lamentable sacrifice of human life which every gale causes, without an ardent desire to see improvement. We who *think* ourselves well posted up in meteorology, whether with official authority or as lunarists or otherwise, would do well to sink all differences and prejudices for the promotion of the one object. As a very zealous member of the Meteorological Society, Mr. Thrustans, said in a letter to 'The Times' of

January 27th, 1863: 'It is most unseemly for one man to ridicule another for his praiseworthy attempts to save the lives and property of his fellow creatures'—a sentiment in which I most heartily concur.

Now in order that the question of weather (as it stands in its disconnection with my own deductions) may be correctly estimated, let me, in justice to Admiral Fitz Roy, remark that he does not pretend to *predict* weather, but only 'forecasts' for a day or two as to *probabilities*, while all his telegrams are from *actual knowledge* of approaching danger. With regard to 'forecasts,' therefore, fishermen say to me: 'But what can Admiral Fitz Roy know more than we do? Bad weather never comes out of a fine sky, and depend upon it we have our eyes open.' Alas! many an eye has been closed in premature death from this indifference and self-reliance within the past month.

It is almost impossible, however, for the public to draw other than the following conclusion. Our ignorant or rather uneducated and unscientific fishermen vie with the educated and experienced men of science in asserting their equal ability to 'forecast' weather. I am much struck with the sagacity with which the former collect in numbers at this post for shelter, where no telegrams, no warnings beyond their own almost intuitive sense of danger prompts them. I have often asked them to what warning they yielded? The reply is generally to the following effect: 'Well, there, you see, we are so used to it, that we don't like the looks of the weather, and so we come in.' (I must say, seldom without occasion.)

We, then, are evidently in this position, all the voluminous observations of past years (and again I must urge it), have not advanced the question of *theoretical* warnings, no, not 'one bit,' we have not advanced 'one jot' beyond empiricism. Not all the resources of Europe have furnished one single facility (unless my discovery prove itself to be such) for foretelling weather. We, as a nation, highest in scientific and maritime renown, as in wealth, still feel the humiliating truth that the labours in meteorology of our greatest and wisest philosophers, whatever they may have done for the *theory* of atmospheric currents, &c., have produced no results which can be *practically* available to sailors in forewarning as to weather. Engineering has given our country a network of railroads, and telegraphy has in consequence furnished us, as it were, with *telescopes* for the mind's eye. Our vision has virtually been thus extended, so that instead of our being limited to observations within a district of twenty or thirty miles, the electric wire acquaints the head of the meteorological department in London, as to 'prognostics' hundreds of miles distant, and right good use he occasionally makes of them. But yet this is merely an advance in *mechanical* facilities, IT IS NOT SCIENTIFIC PROGRESS. I can state without fear of contradiction that notwithstanding differences of opinion, on some points, *my sympathies are entirely with the distinguished Admiral and his work of humanity;* perhaps the more sincerely so from my perfect independence of him and of all others.

There is a painful connection between his department and the twin-sister and noble institutions which unite in one common benevolence, the one to save lives, the other to assist them when safely landed from shipwreck. The greater the success of Admiral Fitz Roy in warning, the more is relief afforded to both the National Life Boat Institution and the Shipwrecked Mariners' Society. *Assist therefore the Admiral, and you assist them both in their labour of love and mercy.* Within the last few weeks he, in 'The Times,' asked the British senate to aid his efforts by increasing the parliamentary grant to his well-managed department by a paltry extra fifteen or sixteen hundred pounds. It scarcely will be excused that I presume to advise, but perhaps no man has personally experienced during some early years of his life more of the rough service of rescuing the shipwrecked than myself (I only mention this in justification of touching upon this side of the question). If I have any claim to a hearing I would avail myself of it by saying, fetter not the hands of Admiral Fitz Roy just now in his heartfelt energies; let not an insignificant 'thousand or two' of Britain's wealth stand between him and progress. I can vouch for there being much to be done, *and he is just the man to do it.* Cramp the ardour of the experienced head of this department, and what do you do? You stave off till a later date the perfection of a system almost within your grasp, one which must soon tend successfully towards saving annually much valuable life and merchandise.

Why should not each of the above referred to Institutions receive public assistance from the government of the country? Why consign such valuable societies to the precarious and fluctuating aids and sympathies of the public in their subscriptions?

After a gale, let any one watch the doorway steps of the Hibernia Chambers, London Bridge, south, and notice the occasional crowds (such as I have witnessed) of shipwrecked sailors, forlorn and shivering, and otherwise houseless and friendless, and destitute, who beset the benevolent-minded secretaries for—for what?—actually in hunger for a meal, for shelter, for decent raiment, and (a blessing on such a society!) for conveyance to the bosoms of their families, where wives, and sisters, and children are sighing to embrace their loved ones once more in safety.

Now, picture to yourselves the importance of the official labours of the man whose duty is so to warn the coast as to (if possible) alleviate such distress, to *prevent* such destitution—*who would willingly impede such an officer, or even vex him if it were to be avoided?* It may be, and I fear it is presumption in me to touch on this, but I would like to meet in a fair field of argument him who could have the heart to record a parliamentary vote against such a grant as an aid to shipwrecked mariners, *through their official guardian* the naval-head of the Board of Trade.

CHAPTER VII.

PROFESSOR DOVÉ — MR. BELVILLE — MR. HARTNUP — CAUSES OF DIFFERENCES OF OPINIONS EXPLAINED — MR. C. CHAMBERS — EVIDENCE OF MOON'S INFLUENCE SUBSTANTIATED.

BUT to return to the question of the discovery of Lunar influences on weather as truth or fiction—as a fallacy, or as a boon about to be conferred on science and on the seafaring community with regard to increased safety of life, and on the mercantile community as regards the protection of the millions of their property afloat.

I have great hopes that when Professor Dové recognises my weather system (as yet but in its infancy), his most elaborate and interesting researches into winds and general meteorology will be still more available in explaining weather changes. He also, I am induced to believe, has been impeded by the uncertainties which have beset barometer readings, in want of appreciation of their just value. The undetected influence which seemed to neutralise the usual action of the mercurial column, must, as in the case of the learned Dutch Professor, have occasioned perplexity.

When tabulations from barometer readings are relied upon, the want of a *complete* appreciation of *all* the causes of atmospheric disturbance *must* derogate from the value of conclusions. No one for a moment can dispute the immense benefit which such distinguished men can confer upon meteorology when their powerful minds are directed towards it—and yet, in the fable, the *lion* was once set free by the *mouse*.

If Professor Dové will compare the dates of storms which he quotes in his highly interesting work with my theory, and refer his dates to my Lunar periods, it will probably surprise him to find that so large a majority of his storms occurred on dates conforming to my theories. His elucidation of cyclones must impress every sailor favourably, as his explanations of conflicting aërial currents must convince and instruct every meteorologist; but still there exists the want of that which alone can be available to the navigator, viz. a simple, well-understood rule, or set of rules (say a system), by which *wherever he may be, he may be able to form some reasonable estimate of the probability of storm or quiet weather.* Give him but this and you alter his position. Such has never yet been attained by others; let us by examination of what follows see if I have found the key to it.

Two modes of proof are open to me. I may either take the observations of others and compare them with my Lunar periods, or I can use my own. But as all persons are liable to nurse and dress up to the best advantage the offspring of their own brains, it will better satisfy the disinterested enquirer if I first examine the work produced by those who are not 'Lunarists,' and who *think they have proved* that the moon exerts no *appreciable* influence upon our atmosphere.

As accuracy in the observer, and truth in his statements, must be above all suspicion, I will take the works of well-known public astronomers, and first refer to Mr. John Henry Belville, of the Royal Observatory, Greenwich, as proving that whatever other influence the moon might have upon weather, the barometer (as I have so strongly urged) is in its movements *in a state of independence of the moon's influence—in other words, that even if we can prove the moon to be a disturber of weather, it does not follow that the mercurial column feels or indicates such power.* But the proof that the moon does not affect the barometer is *no proof of her having no power in disturbing the atmosphere.*

In a handy little work published by Mr. Belville (1849), he gives two tables: the first is thus headed, 'A Table showing the age, declination, and position of the moon in her orbit, in some of the most remarkable *elevations* of the barometer.' The table ranges in dates from 1816 to 1849—say 34 years—and we find that of the periods of *highest* barometer:

3 such heights occurred when the moon was in apogee
6 „ „ „ perigee
and 3 „ „ „ quadrature.

The second table is headed, 'A Table showing the age, declination, and position of the moon in her orbit, in some of the most remarkable *depressions* of the barometer.' The table ranges over a period from 1814 to 1848—say 35 years. We find in this that of the periods of *lowest* barometer:

4 such depressions occurred when the moon was in apogee
6 „ „ „ perigee
2 „ „ „ quadrature.

He was therefore fully justified in saying, 'From these tables it *does not appear that the moon exerts any influence on the extreme movements of the barometer.*'

But, with all deference to Sir John Herschel, I would submit that a glance at the above tables warrants the suspicions which have so long attached to *certain positions* of the moon. He says in 'Good Words,' p. 58:

'Others, again, pressed into the service the great and recondite names of *Apogee* and *Perigee*, and professed to determine the character of the lunation from her proximity at new or full to these mysterious points of her orbit. Both the one and the other rule *utterly* break down when brought to the tests of long-continued and registered experience.'

I do not altogether dispute it, but 'perigee' *is* nevertheless conspicuous (*perhaps* accidentally so) in the above tables—indeed, so much so as to warrant further investigation.

My opinions as to the intensity of lunar influence on weather were formed solely on my own long and careful observations; I have, however, been within these few days put in possession of some perfectly independent investigations, which will, I think, settle this part of the question for ever.

A gentleman in the north, personally unknown to me, who, while a member of the Meteorological Society, holds or held opinions contrary to my own on subjects of weather, and was the writer of the letter to 'The Times,' quoted at page 42 in this book, and who has done me the favour of occasional correspondence upon the subject of my supposed discovery, has called my attention to a report to the Marine Committee of Liverpool, recently issued by Mr. Hartnup, the well-known able and eminent astronomer of the Liverpool Observatory, from its particular bearing on the question of the moon's influence.

Mr. Hartnup herein, with his usual clearness, has produced a list of all the gales which have occurred at Liverpool since January 1st, 1852, all such as exceeded in pressure 20 lbs. to the square foot, with the times of full moon, new moon, and perigee placed against each date. From the importance of this paper, as proving some things and refuting others (to which it will be necessary to refer in detail), I insert a copy, which follows:—

GALES OF WIND IN WHICH THE PRESSURE HAS EXCEEDED TWENTY POUNDS ON THE SQUARE FOOT.

Date of Gale	Greatest Pressure on the Sq. Foot in lbs.	Duration of Gale in Hours	Average hourly Velocity in Miles	Direction of Wind	Moon—Dates of		
					Full	New	Perigee
1852.—Jan. 4.	28	5	51	WNW	6	20	10
,, ,, 9.	29	6	56	WNW	—	—	—
,, Feb. 16.	22	1	50	WNW	5	19	7
,, Dec. 25.	42	5	61	WSW	26	10	9
,, ,, 27.	42	10	63	SW	—	—	
1853.—Feb. 26.	33	5	57	WNW	23	7	{1, 26}
,, April 1.	23	1	51	SW	23	7	23
,, Sept. 25 and 26	37	15	59	NNW	16	2	8
1854.—Jan. 20.	22	less than one hour		WSW	13	28	26
,, ,, 26.	43	2	52	W	—	—	—
,, Feb. 8.	21	less than one hour		WNW	12	26	23
,, ,, 17.	27	4	52	NW	—	—	—
,, ,, 18.	31	4	52	WNW	—	—	—
,, Oct. 22.	24	less than one hour		NW	5	21	{1, 26}
,, Dec. 3.	25	less than one hour		WNW	4	19	20
,, ,, 22.	27	less than one hour		WNW	—	—	—
1855.—Apr. 10.	24	3	50	WNW	2	16	12
1857.—Mar. 14.	24	5	54	WSW	10	25	26
1858.—July 25.	27	5	53	WNW	25	10	9
,, Dec. 24.	25	2	51	WNW	20	4	20
1859.—Jan. 25 and 26	22	2	53	WSW	18	3	18
,, Mar. 7.	22	1	53	WSW	18	4	15
,, ,, 8.	22	1	52	NW	18	4	15
,, ,, 15.	22	2	51	NW	—	—	—
,, ,, 17.	21	2	52	WSW	—	—	—
,, Oct. 26.	28	3	55	WNW	11	25	22
,, Dec. 30.	23	1	50	W	9	23	12
1860.—Jan. 22.	22	7	55	WNW	8	22	9
,, Feb. 27.	22	3	53	WNW	6	21	7
,, ,, 28.	24	2	61	NW	—	—	—
,, Aug. 26.	21	less than one hour		W	{1, 30}	16	17·
,, Oct. 3.	21	less than one hour		WNW	29	14	13
,, ,, 9.	24	less than one hour		WNW	—	—	—
1861.—Feb. 21.	24	1	53	WSW	24	9	26
,, Mar. 3.	22	less than one hour		WSW	26	11	26
,, ,, 6.	26	1	54	WSW	—	—	—
,, ,, 11.	21	less than one hour		W	—	—	—

GALES OF WIND &c.—continued.

DATE OF GALE	Greatest Pressure on the Sq. Foot in lbs.	Duration of Gale in Hours	Average hourly Velocity in Miles	Direction of Wind	MOON—Dates of		
					Full	New	Perigee
1861.—Nov. 11 .	22	1	51	SW	17	2	{ 2 / 30 }
,, Dec. 1 .	22	1	56	WSW	16	1	29
1862.—Oct. 19 and 20	27	3	54	W	7	22	24
,, Oct. 21 .	26	less than one hour		W	—	—	—
,, ,, 22 .	22	less than one hour		W	—	—	—
,, ,, 23 .	21	less than one hour		WSW	—	—	—
,, Dec. 19 and 20	29	6	54	{ WSW / WNW }	5	20	20
1863.—Jan. 19 and 20	36	15	54	{ WSW / WNW }	4	19	18
,, Jan. 26 .	26	1	50	SSW	—	—	—
,, Feb. 4 .	21	less than one hour		WSW	3	17	15

Sufficient data are given in the preceding table to test the question as to the *popular notions* of what is lunar influence, but it brings more 'noses' to the 'grindstone' than were expected.

Mr. Thrustans, in the most delicate and most gentlemanly manner, furnished me with his opinion upon this table. He seems to be an able and zealous observer, and in earnest upon the subject of weather warning. In his letter to 'The Times' (referred to) he added, what shows him to be a generous opponent, if one at all, 'Admiral Fitz Roy is endeavouring to utilise the science of meteorology, and, whether he succeed or not, deserves the sympathy and esteem of all really benevolent men.'

Mr. Thrustans' analysis of the preceding table is, I think, very conclusive in one direction. It was not calculated to raise a hope for my ultimate success.

He finds that—

  3 gales occurred at or near new moon in perigee
  2 ,, ,, new moon not in perigee

2 gales occurred at or near full moon not in perigee
3   ,,   ,,   moon in perigee, but neither full nor new
16   ,,   ,,   moon neither full, new, nor in perigee.

and he notes that 'those of one hour and less are not taken into account, but if they had been they would have been rather *adverse* to lunar influence than otherwise.' He adds, moreover, a note of considerable interest, viz. 'that the gales vary between NNW. and SSW., *none from any point between N. and S. by the East.*' Now this valuable report from Mr. Hartnup must as it stands have operated widely and strongly against my 'Weather System,' said to depend upon 'lunar influences;' but Mr. Hartnup, not being aware of what I have styled 'lunar influences,' has omitted, very naturally, one very important element in my so-called theory, hence the impression of Mr. Thrustans.

Let us supply it, and we then read as under:—

I have thought it convenient to separate the gales into lists according to their duration, and have placed against each what I call the 'pith' of my system, viz. the moon's position *in declination*. (N.B. 'L. Eqx.' means, from want of a better term, that the moon was on the earth's equator, and 'L. col. (N. or S.)' signifies the moon's position at her lunar stitial colure north or south.)

I.

List of Gales which lasted above One Hour.

1852. Jan. 4, lasted 5 hours. (I have no register for this period.)

,,   ,,   9,   ,, 6 ,,   L. Col. N. was on 7th, and new moon on 7th.

,, Dec. 25, ,, 5 ⎫
,,   ,, 27, ,, 10 ⎬ ,,   L. Col. N. was on 27th.

TESTS FOR LUNAR THEORY. 53

1853. Feb. 26, lasted 5 hours.	L. Eqx. was on 26th.
1853. Sept. 25, 26, } " 15 "	L. Col. N. " 26th.
1854. Jan. 26, " 2 "	L. Col. S. " 26th.
" Feb. 17, " 4 } " 18, " 4 } "	L. Eqx. " 16th.
1855. Apr. 10, " 3 "	L. Col. S. " 9th.
1857. Mar. 14, " 5 "	Cyclone, by barometer and veering of the wind.
1858. July 25, " 5 "	L. Col. S. was on 23rd; full moon, 26th.
" Dec. 24, " 2 "	Cyclone (see diagram for date).
1859. Jan. 25, 26, } " 2 "	Terrible cyclone.
" Mar. 15, 17, } " 2 "	Distinct cyclone.
" Oct. 26, " 3 "	L. Eqx. was on 23rd, and new moon 26th (see Diagram).
1860. Jan. 22, " 7 "	L. Eqx. was on 19th. Reference to the diagram for the period shows this also to be decidedly the *continuation* of the change of 22nd Jan. 1860. We had it at Sheerness on 21st; my lunar second day after.
1860. Feb. 27, " 3 " " " 28, " 2 "	} Cyclone, by diagram and veering of wind.
1862. Oct. 19, 20, } " 3 "	L. Eqx. was on 20th.
" Dec. 19, 20, } " 6 "	L. Col. S. " 19th.
1863. Jan. 19, 20, } " 15 "	L. Col. S. " 16th; new moon, 19th. This heavy gale *set in* on 16th, as shown by fall of barometer.

Now here are *nineteen* gales, of which—

    13 occurred indisputably at my lunar periods
    5 were cyclones
    1 doubtful

Total .. 19

II.

List of Gales which lasted about One Hour.

1852. Feb. 16	. .	L. Col. S. on 16th.
1853. Apr. 1	. .	L. Col. S. on 31st March.
1859. Mar. 7 ⎫	.	L. Eqx. ,, 5th, P.M. Fall of barometer *commenced* at P.M. of the 6th.
,, ,, 8 ⎭		
,, Dec. 30	. .	L. Eqx. on 30th.
1861. Feb. 21	. .	L. Col. N. on 19th.
,, Mar. 6	. .	L. Col. S. ,, 4th.
,, Nov. 11	. .	L. Eqx. ,, 11th.
,, Dec. 1	. .	L. Col. S. ,, 2nd. New moon on 2nd.
1863. Jan. 26	. .	Cyclone (from reference to barometer).

There are *nine* such gales, of which—

        8 are distinctly lunar
        1 a cyclone
Total . . . 9

III.

List of Gales which were of less than One Hour's duration.

1854. Jan. 20	. .	L. Eqx. on 20th.
,, Feb. 8	. .	L. Col. N. on 9th.
,, Oct. 22	. .	L. Eqx. on 20th (new moon 21st).
,, Dec. 3	. .	(No diagram to refer to).
,, ,, 22	. .	L. Col. S. on 20th.
1860. Aug. 26	. .	L. Col. S. ,, 25th.
,, Oct. 23	. .	H.M.S. 'Porcupine's' cyclone.
,, ,, 9	. .	L. Col. N. on 6th. (The change began on the 6th, and *increased* to 9th).
1861. Mar. 3	. .	L. Col. S. on 4th.
,, ,, 11	. .	L. Eqx. ,, 11th.
1862. Oct. 21 ⎫		
,, ,, 22 ⎬ .	.	L. Eqx. on 20th (new moon 23rd).
,, ,, 23 ⎭		
1863. Feb. 4	. .	L. Eqx. 6th (doubtful).

There are *twelve* such gales, of which—

              9 are distinctly lunar
              1 a cyclone
              1 (no records to refer to)
              1 doubtful
    Total . . . 12

Now I have long declared that my lunar system is one of universal application (I have yet to prove this); if so, we should be rather inclined to look for predicted *effects* of lunar action among records of storms of longest duration. We should also be inclined to suppose that gales of less than one hour's duration would be attributable rather to local causes, as Mr. Thrustans seems also to expect, than to one general comprehensive system. And yet I find from the above that the same 'lunar influence' must be held responsible for all which occur at certain periods (I have yet to show that cyclones are inclusive).

Those accustomed to *see* the projected curve of the barometer risings and fallings, will understand me when I say that *a gale may happen on one day, which is the result of a disturbance manifested by our instruments as long as two days previously.* The continuing fall of barometer and increase of effects lead to the height of the storm; these must be taken into account, and *reasonably too*, when credence is given to my assertions as to lunar periods in the above abstracts: the second day in a lunar period is marked distinctly. I can only say that I am ready publicly to maintain in detail the accuracy of what I show as regards the above-recorded lunar periods. With this reservation, however (for I am asserting an enormously important fact, and need be explicit), I think Mr. Hartnup has used the *astronomical day in his table*, while, from want of Ephemerides for years antecedent to 1860, I cannot (unassisted) refer to the *precise hour* of stitial colure or L. Eqx. in the earlier dates. This, however, is a question of some twelve hours, which are not worth consideration when we talk of *periods* of weather changes; and again Captain M'Clintock, hereafter quoted, most probably used the *nautical day*—but this cannot really affect the question. Nevertheless, if my proofs rested on this alone I would be more exact, and procure information; but I have other *quite evident proofs* yet to offer, and therefore, having referred to the above in a spirit of fairness, I proceed. But to recapitulate: of gales of

Above an hour's duration	13 were lunar,	1 not,	and	5 were cyclones
Enduring one hour	8 ,,			1 was cyclone
Less than one hour	9 ,,	2 not		1 ,, cyclone
	30 lunar	3 not		7 cyclones.

But strong as this may appear in my favour, even without my claiming cyclones as due to lunar periods, I am about to adduce other thoroughly independent lists, one of which is from another well known astronomer.

Remember I am taking other men's data, *recorded by them before my theory was even discovered.* There are in these no notices of changes of wind or other evidences which I am in the habit of looking for and noting in the recording of fulfilments of my predictions. No projected course of the rise or fall of mercury to guide me, I have taken simply the crude record of certain facts, made by, as I say, *independent* observers.

Now, in illustration of the disadvantages under which I have laboured, take the memorable Royal Charter gale (memorable to me, for it stranded the big ship in which this is written, at the head of Sheerness pier, the end of which it demolished)—the gale of October 26th, 1859. The lunar equinox happened in this case three days before it. Many would, in consequence, disallow of my claiming a gale of the 26th, as due to disturbances which occurred on the 23rd; but the truth is, the new moon (*only just past her perigee*) happened on or about my *always-cautioned-against* 'second day after,' thus heightening the ordinary disturbance to an extraordinary extent;

while the barometer had really commenced, or rather recommenced, its fall on the precise day of lunar equinox—that is, being interpreted in this case, the day of change *predicted by me long beforehand*! Let us now, as another instance, take another gale mentioned in Mr. Hartnup's list, viz. the terrible storm of October 19th and 20th, 1862. The precise period of lunar equinox happened at 9 A.M. on the 20th; this same gale continued with slightly-abated violence for several days, and why? Simply because the new moon at 8 A.M. was nearly in perigee—the lowest barometer, according to Admiral Fitz Roy's report in his 'Weather Book' (page 444), being *precisely on the* 20*th*, the said *day of lunar equinox*.

But I think the following fully illustrates the main cause of differences of opinion as to the moon's power of disturbance in the earth's atmosphere, AND WILL, I HOPE, SERVE TO RECONCILE CONFLICTING OPINIONS. All meteorologists acknowledge that the moon's attraction *does* affect the question; but it appears that, because such is *not as a rule* felt at her quadratures and syzygies, although it is sometimes coincident, its power of appreciable disturbance is repudiated. We have seen that even Sir John Herschel is of this opinion. Some, for instance, have maintained that the moon's phases are periods of disturbance, and others that they are not.

I have fortunately an abstract from a letter sent by a gentleman at Paraguay to a friend in London in 1852, who has long been an enquirer into meteorology. The receiver of this letter has kindly furnished me with the abstract, as bearing on the present public question of lunar influences on weather.

Indeed, I will not omit this opportunity of returning thanks to the underwriters at Lloyd's for their kindness on all occasions, and in particular to one of them (C. H. Allen, Esq.) for the following abstract, and for the encouragement which his early recognition of my theory gave me, coupled, as it has so often been, with valuable authentic meteorological notes from the registrations of Lloyd's costly instruments. It was hard to have to plod 'up the hill' with, at first, scarcely a sympathising friend, and especially when big stones of prejudice were rolling down, as purposely hurled against me in threats of annihilation. *His name, therefore, who spoke the cheering and refreshing word,* ought to be remembered whenever mine shall, in connection with weather, be hereafter mentioned approvingly. Nor shall I forget the readiness in furnishing abstracts from the above-named registry, which has been shown on all occasions by Mr. Lowe, the managing clerk.

As a totally independent report from a foreign station, the following is interesting and instructive:—

Extract from Letters from Paraguay, by C. B. Mansfield, 1852.

'The sailors on the river assert, in defiance of the European wise men, that the weather is regulated by the phases of the moon, the wind changing at the moon's quarters. Now it is a curious fact that for the last six weeks, ever since we have been on the Paraña, we have had a thunderstorm very nearly at the time of the moon's changes, followed in each case by a shift of the wind from north to south. The days of the moon's changes were September 13 ●, 20 ☽, 28 ○, October 6 ☾, 13 ●, 19 ☽ (1852), and the days of our thunderstorms have been September 14, 23, 27, October 9, 15, 22.'

Now, if we tabulate the above, and add the dates of changes to have been then expected in accordance with my 'Weather System,' we shall see as follows, viz. :—

1852	Dates of Storms	Lunar Days
Sept. 13th new moon .	Sept. 14th .	15th Lunar Eqx.
„ 20th first qr. .	„ 23rd .	21st Lunar Stitial Colure S.
„ 28th full moon .	„ 27th .	28th Lunar Eqx. ;
Oct. 6th last qr. .	Oct. 9th .	6th Lunar Col. N.
„ 13th new moon .	„ 15th .	12th P.M. Lunar Eqx.
„ 19th first qr. .	„ 22nd .	18th P.M. Lunar Col. S.

The dates of the moon's phases, as it happened, *so nearly coincided* with the periods of disturbance incident to the moon's being in certain parts of her orbit, that casual observers, therefore, attributed atmospheric disturbances to the one cause instead of the other. And this has doubtless been the case on many occasions where opinions have been formed or abandoned. To me it is another proof of the stability of my lunar theory. I think little of the thunderstorms, as most probably very often due to local condensations; but the '*shifts of wind*' in each case referred to are significant, if not conclusive.

I could write volumes were it necessary in demonstrating the infallible accuracy of my simple lunar system.

Friends have smiled at my seeming enthusiasm in saying that it is 'infallible.' I have scrupulously preserved to the present time the same words in explanation which I adopted in first publishing the theory, in order that my consistency in first announcing the discovery might not be called in question. I had better here give them as still published for comparison with the earliest ones :—

'List of days on which the weather may reasonably be suspected as *liable to change, most probably towards high winds or lower temperature*, being especially periods of *atmospheric disturbance*.

'*The above apply to all parts of the earth's surface—even* (*in a diminished degree*) *to the trade belts.*

'N.B. If the day marked prove calm and still, distrust the day after, and *especially the second day after*.

'The changes vary in intensity, but even at *quiet* periods they may be plainly traced in the scud flying with a velocity totally at variance with the state of the air at the earth's surface, and the clouds at such times generally have a liny or stratified appearance, which usually indicates approaching rain.'

Who will now declare that the moon has no effect on weather?

But I am in earnest, and will bring another promised 'gun' to bear on old prejudices—in the records published by another well-known astronomer—viz. Mr. Charles Chambers, of the Kew Observatory.

If we take a list, as published in the 'Intellectual Observer,' of what I have called 'lunar periods,' for the year 1863, and against the given dates place the number of registered miles of wind for each day, *as compared with the day before and day after*, we ought, if my theory of changes occurring at such times be correct, to see it in some measure corroborated by the Kew registrations.

Date of Lunar Equinox, or Lunar Stitial Colure	Miles of Wind at Kew		
	Day before	Lunar Day	Day after
1863.—Jan. 2	510	244	450
,, ,, 10	160	512	177
,, ,, 16	363	448	285
,, ,, 23	529	595	530
,, ,, 30	548	648	315
,, Feb. 6	316	287	269
,, ,, 13	142	115	430
,, ,, 19	136	76	167
,, ,, 26	139	309	281
,, Mar. 5 (P.M.)	131	139	454
,, ,, 12	185	295	100 (estimated)
,, ,, 18	303	107	215
,, ,, 25 (8 P.M.)	105	76	288
,, April 2	232	127	86
,, ,, 8	351	297	302
,, ,, 15	155	78	109
,, ,, 22	348	466	345
,, ,, 29	345	339	356
,, May 6	261	159	224
,, ,, 12	330	497	405
,, ,, 19	278	771	267
,, ,, 26	333	140	141
,, June 2	200	208	139
,, ,, 8	365	421	292
,, ,, 15	172	228	202
,, ,, 23	329	123	(no record)
,, ,, 29	199	187	202

Date of Lunar Equinox, or Lunar Stitial Colure	Miles of Wind at Kew		
	Day before	Lunar Day	Day after
1863.—July 5	126	173	318
,, ,, 12 (midnight)	109	110–229	68
,, ,, 20	174	85	175
,, ,, 27	150	78	76
,, Aug. 2	(no record)	529	294
,, ,, 9	332	213	236
,, ,, 16	367	396	270
,, ,, 23	114	334	314
,, ,, 29	160	205	34
,, Sept. 5	215	296	346
,, ,, 13	165	198	116
,, ,, 19	142	407	293
,, ,, 25	162	166	129

By the above we see that the wind is greater on lunar days as compared with the day before and day after in . . . . . . } 20 instances.

The wind is greater on lunar days as compared with previous days alone in . . . . } 29 ,,

That the wind diminishes in . . . . 11 ,,

From this list, therefore, we draw two inferences, viz.

1st. *It does appear that the 'lunar days' are really days of disturbance.*

2nd. *That the changes are in favour of 'high winds,'* as stated in my early and continued communications to the 'Nautical Magazine,' detailing fulfilments.

Now, whatever conviction may have reached the minds of the readers of the preceding pages, my own is, that *nothing more is needed to demonstrate the evidence of the moon's actual and thoroughly appreciable influence on the weather.*

As Sir John Herschel suggests, 'lucky hits' may occasionally be made, but he had never probably tried the question of the moon's declination in his researches. Such corroborations as I have given, even were they only lucky hits, seems to offer sufficient reason why our illustrious heads of science should once more turn their thoughts to the question of the probability of Lunar influences acting on weather *appreciably.*

There is, of course, no importance in the following abstract from a Chepstow paper, written by a Bristol observer, except that it shows the influence which the dictum of public officers has upon the opinions of ordinary men. On January 2, 1864, it says (and of course inoffensively):

' I believe that the time will come when the science of meteorology will lead to many more practical results of a useful nature than is at present the case, and that careful observation will then be better appreciated than is at present the case; but I do not anticipate that this desirable result will occur in my time, whatever *lucky hit* may occasionally be made.'

I cannot forget that I am addressing a British public, and that, in the words of the great Napoleon, we as Englishmen ' *never know when we are beaten.*' It is only therefore by vigorously following up an advantage that we can hope to destroy popular errors. That I now, therefore, bring some of my personal experiences to bear on prejudices as to the moon's influence becomes an absolute necessity.

Be it remembered that such prejudices declare—

1st. *That all opinions of lunar atmospheric disturbances break down when brought to the test of experiment.*

2nd. *That we cannot predict weather for more than a day or two in advance.*

Let the reader cast aside for awhile all that has preceded and *weigh my Weather System by what follows alone,* and he will either declare that I have discovered (if not a ' system ' or a ' theory ') at least a *clue* to what is an immense advance in meteorology, so far as regards predicting weather; *or* that I am one of the most fortunate of men, for I can boast of at least more 'lucky hits' *than any man ever did before me.*

Having thus, I consider, furnished ample testimony in favour of a *fair hearing* for my theory, I will add, as derived from various experiences of others, still further proofs of the care with which I have proceeded, in order to acquaint all who are honestly desirous of seeing progress in weather wisdom, as to my peculiar views.

I am sorry to be obliged to give so many abstracts, but it behoves me rather to overwhelm the subject with testimonies than to leave any loopholes for doubt.

In publishing the following, my letter was accompanied by illustrations for the private satisfaction of the highly talented official editor of the 'Nautical Magazine,' a man of the widest experience. Perhaps we may attribute to those convincing illustrations, Captain Becher's kind and complimentary editorial note in the 'Nautical' of September 1860, as under:—

'Mr. Saxby appears to be in a fair way to establish a claim to the grateful thanks of all who are interested in a fore-knowledge of the weather (and who is not?) by the success which has hitherto attended his warnings on the subject. He has (as he says) in our July number (p. 357) foretold the days on which dirty weather may be expected in the remaining months of this year. Our seamen and fishermen ought to look to it—*Mr. Saxby knows what he is about.*'—ED.

In the 'Nautical' for January 1861 appears the following:—

'Perhaps the most severe testing that it has undergone was one suggested by my son (the Rev. S. H. Saxby) who, naturally jealous of my reputation, adopted the following method:—He took various popular chronologies of recent events variously recorded— in *Hannay's Almanack*, for example. He abstracted with care *each* mention of gales, bad weather, changes, &c., and then turned to the ephemeris for the year given and sought for the moon's declination, &c., at each such period. I beg to enclose a copy, and leave it to be judged whether I ought to let the present generation be deprived of the benefits of my discovery and yield my ground to another. The enclosed is for the year 1858, recording events in 1856, and mentioning some thirty instances of bad or extreme weather.

'Now, this date of 1856 was taken at random, but it refers not to any particular portion of the earth's surface, for it speaks alike of the Atlantic, the British Isles, America, the Straits of Magellan, the Duchy of Saxe Coburg, the Gulf of Mexico, Malta, the Philippine Isles, the Bay of Biscay, &c. So much, then, for a general summary.'

In justice to Captain Becher I append the analysis of Hannay's Chronology for, say, the year 1858. Nothing can be more *independent* than mention of gales from unscientific observers.

LIST OF ALL THE GALES, ETC., RECORDED IN HANNAY'S TABLE OF REMARKABLE OCCURRENCES IN 1856.

1856.			*Lunar periods.*
Jan.	5.	Severe snowstorm in Atlantic, lasted eighteen hours.	L. Col. S. 7th (new moon 7th).
,,	21.	H.M.S. 'Dido' dismasted off the Society Islands.	L. Col. N. 20th.
,,	24.	Terrific '*whirlwind*' at Chiswick, &c.	Distinct cyclone.
Feb.	8.	Fearful gale at the Tyne and many other parts; great damage.	L. Eqx. 9th early A.M.
,,	22.	Fearful storm at Constantinople.	L. Eqx. 23rd early A.M.
Mar.	8.	*After* a dreadful storm a ship sinks in Strait of Magellan.	L. Eqx. 8th A.M.
April	9.	Fearful hurricanes . . .	L. Col. N. 10th.
May	7.	Violent gale: London, &c. . .	L. Col. N. 8th.
,,	15.	Gale	L. Eqx. 15th.
,,	31.	Terrific thunderstorm and hurricane, Duchy of Saxe Coburg.	L. Eqx. 29th.
June	30.	Tremendous storm south of England.	L. Col. N. July 1.
July	15.	Very violent storm lasting hours at London—much damage.	L. Col. S. 15th early A.M

1856.		Lunar periods.
Aug. 7.	Severe storm in Kent	L. Eqx. 5th.
„ 17.	Fearful and destructive storm at Brussels.	
„ 20.	Gale in English Channel	L. Eqx. 18th.
„ 21.	Heavy gale at Hartford, Connecticut—'Charter Oak' blown down.	
„ 31.	R.M. steamer 'Tay' lost in Gulf of Mexico.	L. Eqx. Sept. 1.
Sept. 22.	Whirlwind at Glastonbury—great damage.	L. Col. N. 21st.
„ 28.	Heaviest fall of rain registered for fourteen years.	L. Eqx. 28th.
Oct. 6.	Bristol visited by terrific thunderstorms.	L. Col. S. 6th.
Nov. 4.	Terrific hurricane – Montreal	L. Col. S. 2nd.
„ 9.	Heavy hail or *ice* fall at Malta. Ice contractors collected it in carts.	L. Eqx. 9th.
„ 27.	Five thousand houses in Philippine Isles destroyed in a fearful hurricane.	(Not Lunar) (L. Col. 29th).
Dec. 5.	Heavy gale—Bay of Biscay	L. Eqx. 6th.
„ 19.	Furious tempest on American coast.	L. Eqx. 19th.
„ 25.	Violent storm at Gibraltar	Moon with a *few minutes* of L. Col. S.

Of the above 24 gales, 22 occurred at my Lunar periods, and 2 are doubtful. The 'coincidences' are wonderful, too much so to deserve being called 'lucky hits,' as the hits are not made by me, but by Nature itself.

CHAPTER VIII.

Sir F. Leopold M'Clintock — A Pilot's Testimony — Construction of the Barometer, and its Action.

I NEXT give details of *all* mention of weather made by Captain M'Clintock, in 1857 to 1859, in the Arctic Circle, precisely as was done with Hannay's Chronologies for several years, and with the same result.

Abstract of *all* Special Notices of Weather in Captain M'Clintock's Arctic Narrative, 1857 to 1859.

1857.   *Lunar periods.*

Aug. 7. Gale yesterday, to-day changed to calm, foggy. } not Lunar.

" 15. 16. 17. } Anxiously looked-for N. wind sprung up, &c. } L. Col. N. 15th.

" 30. Gale set in, lasted 48 hours . } L. Col. S. 29th.

Sept. 9 to 13. } Gale, and low barometer, cleared up on 13th. } Not lunar.

" 18. NW. winds set in . . . L. Eqx. 18th.

Oct. 24. Furious gales, and brilliant meteor. } L. Col. S. 23rd.

Nov. 19. Heavy southerly gale . .} L. Col. S. 19th.
" 21. Gale changed to NE. &c. . }

Dec. 4. 'Remarkable paraselena'. . L. Col. N. 2nd.

" 17. 18. } Much aurora . . .} L. Col. S. 16th.

" 28. Symptoms of *coming* gale .} L. Col. N. 30th. (Perigee, 29th.) (Full, 30th.)

1858.		*Lunar periods.*
Mar. 1.	Capt. M'Clintock says the winter temperature may be said to have passed away by *February* 10.	L. Col. S. Feb. 9th.
Feb. 24.	Fearful gale of wind	L. Col. N. 22nd.
Mar. 3-4.	Well-marked revolving storm	L. Eqx. 1st.
,, 23.	*Yesterday,* very heavy SE. gale	L. Col. N. 22nd.
,, 27.	Strong NW. gale, with return of cold weather.	L. Eqx. 28th.
April 4.	Furious gale, very cold	L. Col. S. 5th.
,, 17. 18.	Northerly gale, blows hard	L. Col. N. 18th.
,, 24.	Strong SE. breeze, snow, &c.	L. Eqx. 24th.
May 8.	Gale and heavy sea *fast rising*	L. Eqx. 9th.
,, 14.	'Summer just burst upon us'	L. Col. N. 15th.
,, 21.	Southerly winds set in, and continued for 6 weeks	L. Eqx. 22nd.
June 27.	(Date uncertain)	Date uncertain.
July 3.	Westerly wind blew freshly	L. Eqx. 3rd.
July 30.	Gale, and deluge of rain	L. Eqx. 30th.
Aug. 7-9.	Heavy gale	L. Col. N. 5th.
25 to 28.	Change of wind, moderate on 28th	L. Eqx. 26th.
Sept. 22.	Very vivid flash of lightning (most unusual occurrence).	L. Eqx. 22nd.
Oct. 19.	On 17 to 18 a NW. gale of great violence *abated* on 18th. Capt. M'Clintock says it generally blows a gale of wind here (being then in Ballot's Strait).	L. Eqx. 20th (local).
Nov. 2.	NW. gale set in, lasted for 70 hours. Th. down to 12°.	L. Eqx. 2nd.
1859.		
Feb. 8.	'For the last 4 days strong winds and intense cold'	L. Eqx. 6th.
,, 18.	'The old NW. wind sprung up and continued for *several days, and* the Th. fell to 48°.	L. Eqx. 19th.
,, 22.		
Mar. 18.	It blew a NW. gale	L. Eqx. 18th.

1858.		*Lunar periods.*
Apr. 25 to 27.	Very heavy *Southerly* gale .	Not lunar.
May 7.	'Bad weather has now fairly set in, accompanied by a most intense degree of cold.	L. Col. N. 5th.
„ 13.	'A most furious gale lasted until morning of 16th, and it is snowing for a wager.	L. Eqx. 12th.
Aug. 3.	NE. gale sprung up with two claps of thunder (rarely occurs in these latitudes).	L. Eqx. 2nd.
„ 8.	*Gale from westward* . .	L. Col. S. 8th.
„ 15 to „ 17.	East wind died away, and west wind sprung up into a smart gale, falling calm on 17th.	L. Eqx. 16th.

We have here records of 37 gales and extraordinary weather; of these 33 occurred precisely at Lunar periods, 1 was doubtful, and 3 were decidedly (so far as such meagre data can decide) *not* at Lunar periods.

Now, one would almost naturally have supposed that in a region so remote and so physically peculiar,—locked in nearly perpetual ice, subject to occasional thaws and congelations upon a mighty scale,—the state of the weather would be affected by laws different in some measure from those which govern the sunny zones of lower latitudes. Taking the general public impression as to the effect of electricity in those regions where the aurora gives its splendid evidences of electro-magnetic excitement, we should, from the extraordinary activity which such phenomena seem to denote, have reasonably expected a marked exemption from influences which regulate atmospheric disturbances in other climates.

But no! M'Clintock's narrative confirms the simple truth that *there is one general weather system available for all parts of the earth, acting of course with variable intensity at the same moment at different places.* Any apparent exception to this ought, therefore, to be capable of some elucidation by me; although it should not be demanded that a question which has occupied so many abler heads than mine for so many years, without success, should be solved in all its bearings in a comparative day or two.

But to resume. The navigator may argue that because the system I propose professes to indicate changes of some sort about once a week, his experience in passing through the belts of the Trades is repugnant to my views; because (as I can myself also testify) in entering the Trades, he can, in full confidence of settled weather, and according to the season, bend his oldest suit of sails, stopper his topsail sheets to the yards, &c. To this I would answer that unless the indefatigable Maury be in great error this is a beautiful and cheering evidence of the consistency of my lunar theory; for if the changes of the moon in her declination have power to disturb the electric equilibrium of the two hemispheres, the equatorial regions would be the grand laboratory for the production of the effects of such disturbances.

But Maury demonstrates that there is a continual ascent of air (in whatever state of disturbance or repose it may be in other respects) from the great belt of equatorial calms, and that it is carried *above* the neighbouring strata of the NE. and SE. Trade winds;—that from the higher portions of the atmosphere these descend again, meeting the earth's surface at the calms of Cancer and Capricorn, sweeping the faces alike of the temperate and frigid zones, even to the regions of polar calms themselves. The regions of the Trades, therefore, at the earth's surface ought to be exempt from the direct influences, which, as Maury shows, pass high above; while such influences would be in active operation from the calms of Cancer and Capricorn along the earth's surface to the arctic calms at either pole,— and wide experience in observation seems to confirm these, for the moon's influence at the Trades, *although felt*, is much more feebly so than elsewhere.

And it is again remarkable that while the periods between the times which I mark as suspicious are from five to seven days, the *average* of all the storms in M'Clintock's narrative falls within an hour or two per day of my precise periods of expected change,—the amounts of retardation compensate the amounts of acceleration. Why this is not mere coincidence! it is proof! and as such I beg to be allowed to consider it *until these coincidences can be explained by any other man living by any other system.*

Again, in May 1861, the following was published in the 'Nautical Magazine':—

'On the 8th instant, the warrant officer in charge of H.M. gunboat "Lively" being alongside this ship on duty, acquainted me, in the presence of several persons, with the following:—Mr. Millbank (whom I never saw), a Sheerness pilot, had asked him to take an early opportunity of requesting me to give him an extended list of periods of suspected changes of weather *for his use at sea.* He said he had copied out early in the year such a list from the "Sheerness Guardian" (in which they were published three months since in advance); and Mr. Millbank having been, as he said, at sea during the whole of March, had found very great benefit and convenience from my forewarnings against periods of suspected changes, *and they had been of great service to him as a pilot.* The officer (well known for his high character) further acquainted me that the pilot offered the use of his name (if of public utility, and I think it is), as he declared he had carefully, and, indeed, anxiously watched every such period, and that he had a strong conviction of the value of my discovery, even from his own experience.'

That the pilot is a man of experience may be gathered from the fact of his being an active, highly-intelligent veteran of seventy-five years of age, respected by all who know him—still going to sea in the worst weather when his valuable services are called for. 'His *further* subsequent experience of two and a half years, of my system of weather warning,' has abundantly confirmed his opinion in my favour as above expressed, and hundreds of merchant captains (his friends and acquaintances) have coincided with his opinion.

It may be convenient to some of those who read this to have a few hints as to the principles on which the barometer is constructed, and on its action.

Let us consider that the earth is surrounded by air, to the height of (suppose) 40 miles. Now, were it possible to set up a glass tube of this length, having a sectional area of 1 *square inch*, perpendicularly on any part of the earth's surface, the air required to fill this tube would weigh nearly $14\frac{3}{4}$ lbs.; this then would be an air 'barometer,' and supposing the bottom to be open, graduations on the upper portions of the tube would measure the height or changes in height of the air then resting on the part of the earth at which the tube was erected. But such a barometer is a manifest impossibility, we consequently have recourse to other methods.

Suppose instead of air we use water—in a tube—as was done at the Royal Society in 1831; but the tube was of glass, and forty feet long, and thirty to forty feet is a useless height for public observation were there no other objections; but if we take a heavier fluid, such as mercury, we can form an available instrument in the following manner, one which will at the same time illustrate the nature of atmospheric pressure. If we have a tube bent like the letter 'U,' and place a quantity of mercury therein, the two ends being open, the mercury will stand in the two legs at a dead level; but if we close one leg securely with a plug having a small pipe connecting the air above the mercury in the closed leg with an air pump, every stroke of the latter, which pumps out air, will allow of the mercury rising in the closed leg until it attains the height of about thirty inches; when all the air in the closed leg will have been removed, the vacancy or void thus produced, is what we call a 'vacuum.' The rise of the mercury in the tube was caused by the *weight of air resting on the surface of the mercury in the open tube*, which forced the mercury up the empty one where there existed nothing to resist it. Now, if the sectional area of this tube be one square inch, the outer pressure of air would sustain, at about thirty inches high, a column of mercury weighing (as in the case of the contents of the air barometer) $14\frac{3}{4}$ lbs. This, then, is called the 'barometer.' (Barometers are very easily made without an air pump.)

We have declared the air in a quiescent state upon the earth's surface to weigh $14\frac{3}{4}$ lbs. per square inch; but suppose that fluid air is affected in the same manner as fluid water (and such is the case), undulations would produce differences in the *height* of such air—would therefore produce differences in the *weight* of such air, and we should see the surface of the mercury rise and fall accordingly, only we do not register the changes of *weight* of air or mercury; but, more conveniently the changes in *height* of the mercury. Hence, then, our readings of the barometer, as indicating *rise* and *fall* of mercury.

Any person of ordinary intellect can be taught to use and read off instrumental indications; it is within the power of fishermen to use the barometer; but it is the combination of facts, the leisurable *unprejudiced* contemplation of such facts, and the bringing them to bear upon the accepted notions of mankind as regards weather, that is equally open to the professor or the private individual, except that the former has the advantage of immense resources to aid in his research; nor need we concede that all the ability and experience of the country is concentred in one mind. An unfettered person has perhaps an equal advantage in being able frequently to suggest improvements, while those who are dragged along the 'tram-road' of routine must yield to conventionalities, or at least confine their zeal within the limits prescribed by popular notions. I have been a free and very close observer for many years, and attribute my discovery of principles to the elaborate manner in which I have watched the barometer and the weather (frequently by night as well as by day), and more particularly *to the form in which I have projected it.*

CHAPTER IX.

PLYMOUTH MERCHANTS — SCOTCH SEAMEN — SOMERSETSHIRE AGRICULTURISTS — INDEPENDENT PUBLIC TESTIMONIES.

As another totally independent and particularly gratifying testimony to the working of my weather system, as furnished spontaneously by perfect strangers, I think it may disarm those who doubt the accuracy of my theory if I produce what has already met the eyes of many shipowners and merchants on the coast of Devon (and western coasts of England), where, from the situation of the important port of Plymouth, exposed as it is to the force of the Atlantic waves, changes in the weather are more noticed than in sheltered places.

Not many weeks since I received a folio printed notice from Mr. Jenkyn Thomas of Plymouth (the well-known and respected printer and bookseller of that town, but a total stranger to me), of which the following is a part. I knew nothing of its preparation, and was first acquainted with its existence only by the receipt of a 'proof' kindly asking if I should object to my letters in the 'Standard' (therein reprinted) being made still more public. Like most men who have had a hard scientific struggle, with *everything to lose and nothing to gain*, beyond the possible satisfaction of conferring a boon on posterity, whose applause may honour, but cannot benefit the one most interested, I of course consented to its publication. It set forth that :—

'The object of the publisher of the following extracts, which from time to time have appeared in the public papers, is simply to lay before the maritime locality of the port of Plymouth Mr. Saxby's Weather System for the years 1863–64, and which are stated by him as "The Fourth Year of issue." The importance of this system has been acknowledged by several gentlemen in and around the seaboard of Plymouth, and whose precautions in consequence have been admitted as the means of the *timely protection of much valuable property, as well afloat as in storehouses,* &c. at localities more immediately under their control; *being therefore very anxious for extending the benefit which they have candidly acknowledged to have received*, they were desirous for the unanimous observance of Mr. Saxby's system among the merchants and mariners generally of this extensive and fast increasing port; and it is only upon their recommendation that the subjoined extracts meet the public eye in the present form.'

Now, of all men, merchant captains are not to be deceived in professional matters; they may for a time yield to fanciful theories, but their every-day practices will in the long run set them right in their judgments. Weather is, moreover, with them a question of life or death. Whatever value the reader—the scientific reader—may attach to these pages as evidencing a newly-discovered truth, I shall ever feel grateful to the men of Plymouth for their kind appreciation of my attempts. It honours them and me too. And again, to show the nature of the 'overwhelming additional proofs,' which I could produce if called upon (and it is still necessary for me to adduce others), the reader will, I think, see enough in the following to justify in all 'real meteorologists,' encouragement rather than abuse, misrepresentation, and depreciation, such as it has been my hard lot to endure. The anonymous insulting letters which I have received disgrace the writers, not me.

On the east coast of Scotland mariners use a small work called 'Inglis's Tide Tables' for their ordinary nautical and commercial purposes. It is a sort of handbook for all who sail between the Shetland Isles and Berwick-on-Tweed, and is to the seaman what Hannay's almanack is to the landsman, and is conducted by a most able man at Aberdeen (a stranger to me).

In his Tide Tables for 1864, he says to his hundreds of thousands of readers as follows (at page 9):—

'The warmest thanks of the whole seafaring community are due to Mr. Saxby for these weather predictions, which are now highly valued; *the predictions for* 1863 *in this publication having been so closely fulfilled.* While the present publication is in press, the predictions for October 10th to 12th are fully verified by the heavy storms from east and south which raged from the 10th to the 12th, and during which the Norwegian brig "Wilhelmina Haal" went ashore at Stonehaven, and the "Die Aufgehense Sonne" at Belbevie.'

I can only accept the conclusion to which Mr. Inglis has, from his large intercourse with merchant captains as an extensive ship chandler at Aberdeen, arrived as one of those independent corroborative testimonies which, as an individual unscientific emanation, would have had little weight against the opinions of men eminent in science, but it is, nevertheless, one of the 'stones' by which it seems my structure is destined to be constructed. Hence the necessity for their accumulation at this opportunity.

From a number of letters from experienced naval commanders, I feel called upon to add at least one. A naval officer can scarcely have a more harassing duty than in the conveyance of valuable stores from one naval port to another. To 'knock about' the English Channel in winter time, with perhaps a heavy deck-load of boilers, machinery, &c., is a hard and most anxious duty, and, indeed, *the ablest officers alone can do it with safety.*

One of these gentlemen, who commands H.M.S. 'Fox,'

and who is well known in the service for his great experience, wrote to me as follows:—

'Her Majesty's Ship "Fox," Woolwich: Jan. 2, 1863.

'DEAR SIR,—I am very anxious to have another of your cards; will you kindly send me one for the coming quarter? I have seen your remarks (in the "Standard" of 1st inst.) on the loss of the "Lifeguard," and fully agree with all you have stated therein. I just managed to reach the Downs on the morning of the 20th of December, where I was detained until the 22nd before I could proceed on my voyage. I have on many occasions verified the correctness of your prognostications, and have great faith in them, finding them of great service to me in my present employment.

'Believe me, &c.
(Signed)   'H. PULLEN.'

And in a note in reply to my question if I might publicly use his opinion, he kindly adds: 'I shall be most happy to bear testimony at any time to the correctness of your prognosticating, *having by experience proved it.*'

And further, as regards agriculturists, on February 1st last I received the note copied hereunder from a total stranger at Yeovil:—

'Will you kindly send me one of your Weather Tables, as I am much engaged in barley-farming, and I am anxious to know what kind of harvest we may expect this year.'

My reply was as follows, dated February 2nd, 1863 (at which date I had not published a weather list for the autumn and winter, nor did I until it appeared in the 'Standard' of June 10th last):—'Thank you for your confidence in my Weather System,' &c., and added:—

'July.—Nothing that I can detect seems to indicate disturbance. I hope for a quiet month. If any changes take place, they will probably be on the 5th, 12th, 20th, or 27th.

'August.—Between 14th and 18th I expect weather totally unfit for harvest operations.'

Now, it is curious that an anonymous correspondent (and as such not, of course, worthy of my further notice) sent me a note taunting me with the weather of July last, whereas the very absence of disturbance, except such as would be observed by a scientific observer, was the best fulfilment of my predictions of the previous February 2nd. With regard to the August weather, it is well known to have been a beautiful corroboration of the value of my system, for the weather was as follows, viz.—

August 14—Fresh south-west wind.
       15—Fresh gale from west.
       16—Much rain, and strong gale from west.
       17—Strong wind and heavy squalls of rain from north-west.
       18—Cold, with rain at times.

I ought to add that in one of my letters to Mr. Jenkin Thomas, of Plymouth, in October 1863, I said:—

'Your agricultural friends may, I think, rest assured that we shall have *no long-continued* frost, lasting more than two or three weeks, this winter. The serious disturbances expected in December next, and in January, February, and March would *break up* any such, or greatly interrupt them; while thaws will generally, and almost *as a rule*, occur at my marked periods. I can say nothing of probable intensity of frosts, being as yet only justified in referring to "lower temperatures" at certain times.'

(N.B. Witness the fulfilment of this *frost* and *thaw* in the one which set in on December 31, and was interrupted by the other on January 7 inst.—*the frost setting in at the time of lunar equinox, the thaw breaking it up at the next lunar stitial colure.*)

To show the interest excited by the fulfilments of my

predictions (made, be it remembered, six months in advance), I could refer to applications from persons of all ranks in society, from the general officer down to the humblest fisherman in different parts of Great Britain, for information as to my opinion of coming weather; and although some questions have been too much for me, for I presume not to be a wizard, I could only offer my *reasons* for expecting certain changes at certain times. I *never refused an answer* to a courteous applicant, or withheld one from the poorest, although I could ill afford the time and expense. I do not think the 'general officer' referred to would, if asked, disallow my using his name, but it is not necessary to trouble him (another total stranger). I therefore quote from his letter to show the confidence which the fulfilments of my predictions have inspired in the minds of able men : —

'Having repeatedly found your statements as to changes of weather so very correct, it would be conferring a great favour if you would let .... know when he might reckon on a fine smooth passage from Kingstown to Holyhead. The passage is a short one; but he will be accompanied by a large and young family. This will, I trust, plead my excuse,' &c.

The nature of my answer may be gathered from the following letter of thanks from the gentleman who crossed with his family accordingly : —

'Pray accept my best thanks for your prompt answer to General ———'s letter; but before I had received it I was tempted by the promise of fine weather on the 5th to make a start, and the morning of the 6th being calm, I took advantage of it and crossed. The passage was tolerably good, although there was a fresh (favourable) wind, and rather a swell; *but it was fortunate we came over* then, as *your prognostications have proved correct,* and I now think we are in for a gale from the NE. It is certainly marvellous how *man* has been enabled to foretel the changes of weather *so exactly and so truly*, so far into the future; but you must feel honoured in being the instrument, through God's mercy, of saving many a man from a watery grave. Thanking you again for your kindness,' &c.

Such confidence in my Weather System on the part of the public entails on me considerable responsibility; it has, however, been much alleviated by letters from men of especial ability of judging in such a matter as weather. A gentleman of title in Scotland, a large landed proprietor, a stranger, wrote in July last to me as follows:—

'I have *not* found it' [my theory] 'altogether proving correct here, but I have *no small faith in it*, and hope, by distributing papers on this east, and the west coast, where I have property also, to have it more fairly tested, as I *have* found disturbances at these periods mentioned by you, *small and great*, and believe them to be caused, as you do, by the coincidence of lunar points with the crossing of the earth's equator by the moon, and with her apogee and perigee. *I am an old sailor*, with a retentive memory, and when in the East remember Horsburgh's table, with preface to the first edition of his great work, in which he calculated the chances of changes of weather at the greater from the mean of these coincidences attending each quarter; but I was beginning to doubt all influences of the moon on the weather when I first met with your papers, and believe this followed (in my experience) from not having gone to *any* point beyond the change and the moon's quarters. *I believe you to have made a very important discovery*, for which you will yet receive your reward. *I told my farm manager that if we could cut our hay* on the 5*th*, *I would guarantee him fine weather until the* 12*th, and so it proved*. With my best wishes for your ultimate success and reward,' &c. &c.

Were I not hard pressed it would, for many reasons, be impolitic to use such quotations; but these show the feeling which is operating in many able observers about Great Britain, and which, if adequately seconded, might lead to great results. The citadel I am attacking is a stronghold of prejudices, situated on so commanding an eminence that approach is difficult and dangerous, and I shall succeed or fall in the attempt. I despise, therefore, neither my opponents nor the smallest proffered aid in the struggle.

CHAPTER X.

THE 10TH TO 13TH DECEMBER, 1863 — SUBSEQUENT LUNAR PERIODS.

My notices which warned the public against the great lunar period of 10th to 12th December last have, it appears, found their way to every part of the globe.

In order to add weight to them I frankly explained my reasons for cautioning against what I feared would be the extreme danger of that period, both as regarded heavy gales and extraordinary high tides. The following was published in the 'Nautical Magazine' for March 1863, and also in the 'Standard' on 10th June, 1863 :—

'In order that my confiding friends among your readers may not deem such a prediction or statement presumptuous, permit me to explain.

The new moon will happen in the precise hour in which the moon attains her most southern declination, viz. on the 10th December, 1863, at 8. P.M. Whatever influence the *merely new* moon has generally in raising the tides will be at its height on the second day after. This 'second day after' will be her precise period of perigee, even within three hours. The sun's semidiameter on the date of his perigee (30th December) will be 16' 18·2"; but on the 12th (the second day after new moon) his semidiameter will be 16' 17·2",—therefore he will on the 12th be within 1" of his perigee, as shown by his greatest semidiameter.'

The public seemed struck with the apparent soundness of my views. Shipowners sent, and merchant captains took (spontaneously) my weather lists abroad with a confidence which, however gratifying, has surprised me. I hear from Petersburg, Lisbon, the Cape of Good Hope, Australia, the Coast of Africa, the West Indies, &c., that my Lunar Weather System is as fully believed in abroad as at home—it has worked its own way *as if it had merits.* But the kindness of many observers has led to an overestimation of my ability. It is too much to expect of me that I can define the *terrestrial limits* of any results which at certain times *may* present themselves under lunar influences; and as to what I call Lunar Periods, my list only gives the precise day on which a Lunar Equinox &c. will occur, but whether at morning, noon, or evening is the precise hour I cannot with consistency advertise, because I have for years noticed that many changes occur immediately before, others immediately after such precise times. The atmosphere may, at one period when a disturbing lunar change occurs, be in a certain condition; but at others in a totally different state, therefore different effects must be expected. All that I can do at present (or at least all I at present feel justified in declaring, with some exceptions) is that certain periods named in my lists will be times of *change.* I have found that such changes sometimes set in the day before the one indicated; and when this has been the case the effects have been prolonged till the day after, or more generally till the *second day* after. So that, disregarding all peculiarities of atmospheric condition which may prevail at a lunar period, I, as a rule, consider such 'Lunar Period' in which especial attention to weather is called for, to consist of at the utmost four days, viz. the day before, the day marked, and until the second day after, *for it is during this time in particular that the barometer is not worthy of implicit faith, unless it be falling.*

I am quite aware that I leave only three days in about each seven, in which the chance of settled weather is likely to occur, and this has furnished some rather unscrupulous adversaries with a weapon which has been pointed against me; but will they oblige me by trying their hands in selecting any other regularly recurring periods which will give a uniformity of results like those that are derivable from lunar influences? It must not be forgotten that my lunar system indicates certain periods of intensity, and that many marked periods pass over without any appreciable consequences being apparent to the casual observer; but I am in a position to assert that, to the man of science, they are invariably traceable, even at the comparatively quiet periods of April, May, June, and July.

As regards the 'second day after,' I have not the least doubt that it should in all cases be particularly suspected and guarded against.

I am quite able (thanks to many observers in different localities who have favoured me with copies of their registers) to prove more than I think it necessary to trouble the general reader with at present; for instance, *more rain occurs during my lunar period, taking day for day, than at others.* Local influences may in some cases interfere with this, but taking fairly exposed and open situations the rule indisputably applies. There is very often an increased proportion of rainfall on the second day after.

Now, if the reader will obligingly turn to a preceding page herein, he will see that an *unusual* quiet at any particular district on the 10th to 13th December last was no proof of failure in my system of weather warning; for at page 68 I distinctly say that lunar influences act 'of course with *variable intensity at the same moment at different places.*' I can only, as it were, acquaint the hare beforehand as to *the time when the 'hounds will be out,'* puss must then take her chance of the *direction they may take. It is certain the hounds will not cover the face of the whole country.* It is so with weather: the *changes* I positively predict at defined periods may develope themselves as changes *in various ways and at various places*. It is enough for me at present that I can furnish these periods; I cannot tell where the burst of fury is to fall.

Certain it is that we can draw a line in Europe which separates the places visited by severe storms during 10th to 12th December, from those which were spared; and the southern were actually the quiet portions of the continent. Not every storm which the Admiral (by his instruments) *sees* approaching a district *reaches it*, although he gives the *consistent* and useful warning.

I shall now show that such 'variable intensity' really occurred. Be it remembered my warning in the previous March, was that from the 10th to 13th December would be a period of great danger, and that there would be an *exceedingly high tide* on the 12th.

Not alone in order to vindicate my own integrity in the eyes of the authorities, but I owe it to the tens and hundreds of thousands of English and foreign eager enquirers into the merits of my lunar theory, to *prove the propriety of my having warned the world* as to the 10th to 13th December 1863.

Now, no sooner had the period passed over than I wrote, as I had previously volunteered to do, to the Editor of the 'Standard' newspaper: the following is a copy:—

'I ask, then, the great favour of being allowed to occupy a little of your much-coveted space, while I acquaint your readers, according to promise, with facts as they have occurred in connection with the above warning. And first, as regards the "destructive storm" of which I said there was so much "chance." I know nothing yet beyond what fell under my own observation at Sheerness, and which is as follows:—

'On the 10th we had a most gorgeous and highly-threatening sunset, with *an exceedingly rapid scud from NW.* (after the ominously dead calm of the preceding evening the wind veered northerly).

'On the 11th we had rain and very rapid scud in the forenoon, with very light wind, but rain and strong gusty heavy squalls in the afternoon, and till about three next morning; a number of North Sea smacks running in for shelter towards nightfall.

'On the 12th, exceedingly rapid scud from NW., with strong gusts of wind towards evening; and again many smacks running in for shelter.

'On the 13th, fine but very hazy.

'The sailor will perceive from the above that it was blowing heavily not far from us. (I must leave your readers to draw their own conclusions from what news may yet come from surrounding districts. It is enough for me to prove that the period was one of very great atmospheric disturbance, although we at Sheerness fortunately escaped the tempest.)

'And, secondly, as regards the "most dangerous tide," &c.:—

'In the night tide of the 12th, or rather at 1 h. 56 m. after midnight, the tide rose so excessively that it is a mercy we had nearly a dead calm at the time of high water. According to the royal dockyard tide-gauge register I see that only since 1847 have we had such a visitation, and on that occasion the tide-level was 10 inches less in height than on Sunday early morning. *Had there been a breeze, and especially from north-easterly, the whole of the Sheerness dockyard, with the town of Sheerness, must have been under water, and most destructive would have been the consequences.*

'Already do the daily papers speak of "considerable damage" to the Thames wharfs, the flooding of the Temple Gardens up to "the sun dial," &c.

'Respectfully, then, would I ask whether (with so firm a conviction on my mind in March as to what threatened for December) I did wrong when I fearlessly staked my reputation, and, indeed, official position, in warning sailors and wharfingers, &c., to take precautions? I am most unwilling to trouble you further, but I would like to remark that, whatever bad weather may have occurred during the period I cautioned against, we had no warnings whatever from the barometer, which on the contrary, would rather have misled the generality of sailors; while in the great cyclone of the 2nd inst., every one who possessed a barometer might have seen on the 1st inst. what was threatening.

'Dec. 14th, 1863. S. M. SAXBY.

'P.S. Accounts begin to reach me of exceedingly heavy gales in the north—a "strong gale" at the Goodwin, &c., on the 11th and 12th.'

In further testimony, I add that—

At **Sheerness,** we had the 'most dangerous tide.'

On the **Norfolk coast,** they had also 'a most dangerous tide,' for a total stranger to me, an old sea captain, of Burnham Overy, wrote to me on December 21st, 1863, thus:—

'The tide made his appearance much earlier than usual, at 7h. 45m. (A.M. of Sunday 13th) the tide was at its highest, being a *very large tide. Should we have had a gale from the NW. it would have overflowed all our banks. I think you were perfectly justified in giving warning.* The precautions taken by myself and neighbours will not be removed yet. I may say *your warning has induced a long neglected sea bank to be put in repair.*'

Captain Sturley's first letter to me, dated November 2nd, is a singular proof of the so general belief in my correctness, it was as follows:—

'Observing a letter in the "Standard" . . . . on reading it, am pleased to say your prophecy of the *unharvestlike* weather of the four days in August was *truly fulfilled* in this part of the country, so that I feel great confidence in your calling attention to the coming December 12th, advising those living near the sea to be upon the watch. Last year the heavy gale and tide (to which also I had drawn special attention) nearly came over the sea bank. *Would you still advise us* to take every precaution against the coming tide?'

At **Sunderland,** they had the same exceedingly high tide and high wind.

I have not the pleasure of knowing Mr. John Crosier, of Sunderland, but find he is much respected, and an alderman of the borough. He allows the use of his letter to me of December 22nd last; it is as follows:—

'We had *an exceedingly high tide* and *high wind* here on the 12th; in fact, *I do not think I ever saw a higher.*'

He also adds:—

'I have observed the indications closely, and find as you recently remarked in the " Standard," that the second day after is almost invariably agreeable to indication. But at Jarrow, South Shields, about ten days ago, the weather, as usual, was the subject of conversation, when a gentleman being there, told me had been watching your indications, and *found them remarkably correct* since the beginning of July, so far as regards here with about three exceptions.' (N.B. This verifies *twenty-five out of twenty-eight consecutive warnings*).

At **Haverfordwest,** Pembroke, Mr. Symons, of London, in his return of British rainfall for December 1863, considers this to have been a very stormy month, *particularly from* 10*th to* 15*th*,' and ' very mild, there being no frost.'

At **Boston Spa,** near Tadcaster, Yorkshire, see the ' Leeds Intelligencer' for January 7th, 1864, F. R. Carroll Esq., who gave me permission to use his name as the writer, says the wind was very high.

'December 3rd, the gale was not so severe here as in the south of England: 10th, 11th, and 12th a very remarkable wind—*do not recollect so much wind with so high a barometer: on the night of the* 11*th it blew hard*, and in the middle of the day on the 12th *it blew harder*, with a rising barometer 30·16 in. *The barometer did not foretell this gale.*'

At the **Shetland Isles,** Unst, they had a very high tide on the 12th.

My second son, Henry L. Saxby, M.D., in his usual report to me, says as under, as regards my period of 10th to 13th December.

' Thursday 10th.—Heavy squalls from NW. with sleet and hail.

'Friday 11th.—Cloudy, light NE. wind.

'Saturday 12th.—*Very high tide. Heavy wind* from NNW. from 2 till 11 A.M., after which sleet and rain, succeeded by a *sudden fall* of wind, but heavy squalls, with rain and sleet, were afterwards frequent and *sudden*.

'Sunday 13th.—Nearly calm.'

At **Plymouth,** the NW. wind kept back an otherwise threatening tide.

Mr. S. Phillips, to whom I am indebted for much care in watching, for a considerable time past, the truth or fallacy of my Weather System, and who from his well-known experience and opportunities as Dock Master of Millbay, Plymouth, can give most reliable testimony, wrote to a friend of his (both being strangers to me), who kindly forwarded the letter to me for public use; he says thus (letter dated December 15th, 1863):—

'The rise at our gates on Saturday evening (December 12th) was 16 feet; *we should have had a fearful tide had the wind blown from the southward and westward; what little wind we had was from the NW., which keeps them back in this part of the world.* . . . for the last four days vessels that left Plymouth on Saturday and Sunday returned, some say that blowing hard from the westward in the Channel on Saturday (12th) was the cause of their return; but Sunday and yesterday there was no wind. The Dock Master of the Victoria Dock, **London,** wrote to me yesterday, he says they had *an extraordinary high tide there*, it enabled him to dock the iron clad "Monitor," also the largest merchant ship afloat, "The Great Republic" — He had 31 feet of water at the entrance,' &c.

He also says in a letter, November 13th, 1863, to the same friend :—

'You may depend upon my giving circulation to his (Mr. Saxby's) Weather System. I have it constantly before me and consult it, as the best I have ever seen yet—no office connected with shipping should be without one.'

At **Chepstow,** and of course in the Bristol Channel,

the tide was not high, as thus explained by a letter from an able meteorologist, J. G. Wood, Esq.

I am deeply indebted for the great pains this gentleman has occasionally taken to give me information on such points as could facilitate the *testing* of my 'Weather System.' *I know not whether he be or be not* a 'lunarist,' but he, a stranger, has assisted me considerably, and I beg to thank him. In a communication from him early in January, he says, that on December 13th, the weather was very fine at Chepstow, and adds—

'But I hear that at Bath and Bristol *was a severe storm*, and *that the Welsh coast suffered severely*,' and adds, 'the tides in the Wye were not extraordinary, due probably to NW. winds, which tend to diminish them here, whilst SW. winds increase considerably the height. As to the *gale* we may have been in the centre of it, and so escaped its effects, for it *seems to have been general in many parts of the west country*.'

At **Edinburgh,** as to the wind and tide, a friend writing to a member of my family, says:—

'*The tide was higher than ever I saw it, and it blew like anything* (on December 12th, 13th).'

At the **Orkneys,** the 'Orkney Herald' says:—

'*The prophecies* for December 12th have been fulfilled to the letter; *there was a heavy gale and a very high tide*.'

At **Nairn,** according to Admiral Fitz Roy's own meteorological report, it *blew almost a hurricane* on December 11th, and a heavy gale on the 12th.

At **Cape Clear,** by Admiral Fitz Roy also, it blew a strong gale December 11th.

At **Liverpool,** by Admiral Fitz Roy also, it blew a strong gale on December 11th.

At **London,** by Admiral Fitz Roy also, it blew a strong gale on the 11th.

At **Yarmouth,** by Admiral Fitz Roy also, it blew a strong gale on the 11th and 13th.

At **Aberdeen,** by Admiral Fitz Roy also, it blew a strong gale on the 12th and 13th.

I have little opportunity of acquiring other knowledge as to the weather at the period referred to, but beg to heartily thank strangers and friends for so much *spontaneous* assistance in furnishing reports of weather, to publish *all* which, would greatly and unnecessarily increase the size of this book; enough has been said to enable the public to judge of *the grounds on which* I founded my warnings nine months beforehand, and to examine proofs of their *entire* fulfilment.

The prominence given to my special warnings, demands (under circumstances) every effort on my part to lay the actual occurrences openly before the world, lest I be again misrepresented.

Having, therefore, described lunar influences as affecting weather—as a law and a principle acting in all parts of the world at certain periods—which can as well be named years as days beforehand, I now turn from our own continent, and adduce testimony from almost the Antipodes. (By-the-bye, it is highly important to note that in this instance the moon and sun were in *extreme Southern declination*.)

The undermentioned painfully interesting letter is taken from 'The Times' of 13th February 1864, and is from their 'Own Correspondent' at **Melbourne,** dated 26th December 1863. (I give a large abstract, because in the 'Nautical Magazine' for March 1863, in warning against this expected terrible weather, I used these words, as I did also in the 'Standard' of 1st June 1863 :—

'Now let any man tell me *what other influence can be adduced to coincide for that period* (December 10th to 13th) so as to increase the *chance* of the *most destructive storm* and *the most dangerous tide* with which this earth without miracle can be visited.')

'FROM OUR OWN CORRESPONDENT.

'MELBOURNE: December 26, 1863.

'Such scenes as for the last ten days we have witnessed in the neighbourhood and suburbs of this city are without precedent in the memory of black or white inhabitant. On Monday, the 14th inst., the weather, which had been for some days previously somewhat unsettled, culminated in one of the fiercest and most prolonged gales of wind, at irregular intervals rising to the strength of a hurricane, ever known along the Australian coast. Accompanying the wind was such a deluge of rain that speedily several of our lower streets seemed converted into rivers, and the river Yarra, swollen by the unusual contributions from the Dandenong range of hills, in an incredibly short period overflowed its banks, converted a large portion of our suburbs and of the southern side of the stream into a vast lake, and drove the inhabitants, most of whom were of the poorer class, with precipitation from their houses. The Peninsular and Oriental Mail steamship "Bombay," which arrived off our port as early as the 14th, could neither enter the harbour, nor could pilots get to her, such was the fury of the wind and the impossibility of those on board getting a proper observation of the lay of the land through the thickness of the rain. She therefore lay off Cape Otway until the 15th, when, getting a glimpse of the Heads during a short lull of the storm, she bore up for the harbour and entered safely on the evening of that day. "The Great Britain" came in on the following day. Meanwhile the storm raged on with undiminished fury. All the lower lying banks of the river were overflowed to the height of some forty feet. Melbourne became surrounded by water. Boats plied at first-floor windows. Some of the poorer suburbs, their streets turned into canals, looked like a mixture of Venice and St. Giles's. All navigation of the Yarra became suspended by reason of the impossibility of steamers and other vessels keeping within the river course, now effectually merged in a wide waste of waters. The only indication of the actual course of the stream was the curious succession of objects carried along on its surface—chairs, tables, fowls, fragments of wooden houses, children's cradles, crinoline, four-posters, files of papers, and hundreds of other objects which had been floated out of different tenements and pursued each other down the river in the direction of the bay. Almost all the factories on the banks of the stream were suspended, merely the tops of buildings indicating their situation to the eye. Many market-gardens were ruined. The Melbourne Gasworks were under water, and for several nights the city and a large portion of the suburbs were reduced to candles. Vessels of small draught were here and there lifted out of the river and placed on the river banks. The Hobson's Bay Railway trains ceased running, and for some time a very valuable portion of their property was in imminent danger. The faces of directors looked nearly as long as their own trains. Business for a day or two was almost entirely suspended, and between Melbourne, Sandridge, and Emerald Hill could only be carried on in boats. Hundreds of people came in from the country to behold the sight. For hours they would occupy any elevated position, looking forth wistfully at the certainly somewhat impressive scene. Readers of Dr. Cumming and of the prophetic order of literature shook their heads. Some seemed to think that Dr. Colenso was in some sort responsible, and one preaching man, pointing to the flood, asked me, "What could we expect?" I said I expected the flood to go down again in a few days, an expectation which has become realised, as the river is fast resuming its usual appearance. The Government have opened our now very comfortable and extensive Immigrants' Home to the houseless and the distressed. Subscriptions are being raised, and it is not improbable that a sum of money will be voted in the next session for the same purpose. It is said that something like a quarter of a million-worth of property is destroyed, but this can be little better than a wild guess, seeing that as yet we have had neither time nor opportunity to take stock of the extent of the damage. The damage is at any rate very great, and only time and enquiry can put us in possession of even an approximation to the truth as to the amount of property injured and destroyed.

'Mr.* Saxby, we are told, predicted all this very many months ago. If so, be it a fortunate guess or a meteorological calculation, the result is almost equally surprising. His predictions,

---

* The writer by mistake said *Lieutenant*.

besides the fulfilled past, gave the 23rd and the 31st of December as days on which we might expect unusual weather; and certainly on the afternoon of the 23rd we had a very severe thunderstorm, accompanied by another unusually heavy fall of rain, which, however, only lasted a few hours.'

On the 19th February 1864, I received a letter from my third son, dated **Tasmania,** 14th December 1863 (he was on a visit to a friend at Oyster Bay, on the east coast of Tasmania), acquainting me that so terrible a storm was then raging, that it exceeded in fury all that the oldest colonist in the district had before witnessed. He said the river (the Meredith or Swan?) had, while he was writing, risen twelve feet above its banks, and that the mountain torrents were acquiring a frightful magnitude, to the stoppage of all traffic, remarking that it is 'not every straw or two in the road that will stop a Tasmanian in his journeying,' but that then travelling was an *impossibility*.

This accords in date (14th December) precisely with the bursting of the same storm at Melbourne, a place about 300 miles distant! and says much for the magnitude (as predicted by me) of the atmospheric disturbance at that time prevailing, *occurring moreover at the very height of an Australian summer*!

My son further corroborates the account of the Melbourne 'Times' correspondent' in describing the unsettled weather and 'huge black clouds' of *the few days previous.*

Now I have the honour of laying before the British public the above fulfilment, as to *one* of my lunar periods, from observations in *both* hemispheres.

I now again request the favour of the reader's dismissing from his mind any conclusion or decision to which the preceding pages may have led him. Let us imagine that a weather system is proposed to us for scrutiny, the *merits of which are to rest upon what follows alone*; that the reader is to act according to evidence contained in the succeeding statements only, and is not to be influenced by either the opinions of others, or popular prejudices; but is to take a set of my *consecutive* predictions, and compare them with such weather as is *known to have actually occurred.*

Now, in June last, it is well known I issued the following list of dates, as expected by me to be periods of atmospheric disturbance and change, '*most probably towards high winds,*' &c. (see page 59).

Having already examined with some minuteness the period of 10th to 15th December, 1863, let us attend to *each lunar period that has passed since that date*, even to the latest moment of going to press. The list will be as under:—

December 17th—23rd—31st.
January, 1864, 7th to 11th—13th—20th—27th.
February 2nd—7th to 9th—16th—23rd and 24th.
March 2nd—7th to 10th—14th—22nd

(The statement of actual weather is mainly taken from my own public register; when otherwise, it will be particularised. It is in the power of any one to examine into the correctness of this from Admiral Fitz Roy's interesting and useful daily list in 'The Times,' &c.)

1863.
Dec. 17th.
> North-east gale, with remarkable depression in the barometer. An immense number of small craft from the North Sea sought shelter in the Medway.
> At Chepstow, wind changed from W. to NE.

„ 23rd.
> WNW. Strong gale.
> At Melbourne (Australia), strong gale (see 'Times' letter of 13th February).
> At Chepstow, cloudy, most of the day, with *stormy appearances.*

„ 31st.
> On 30th, A.M. wind NW. and falling; P.M. dead calm. Late P.M. wind Northerly.

1863.
Dec. 31st.
>On 31st, strong gale SSE. to ESE. *Hard frost set in.*
>On 1st Jan. strong gale from E. to S. Great number of smacks sought shelter in the Medway.

1864.
Jan. 7th to 11th (a period of *special* warning).
>On 6th, weather fell dead calm. The intense frost which had set in on 31st continued in its severity.
>On 7th, wind NE. and variable. A *thaw* set in. Dead calm.
>On 8th, variable. Dead calm.
>On 9th, wind shifted to SE.
>On 10th, very high tide (as predicted).

At the Shetland Isles—
>Jan. 8th, wind rising from SW.
>„ 9th, heavy gale ESE.
>„ 10th, heavy gale SSE.
>„ 11th, A.M. moderate and light; from 11 A.M. steadily increasing to a gale. Very high tide (as predicted).
>„ 12th, gale from S. till 11 A.M. Wind rapidly fell. Also very high tide.

On the Yorkshire coast (see 'Yorkshire Gazette,' 16th January)—
>Jan. 11th, 'simultaneously with Admiral Fitz Roy's telegram, strong breeze sprung up and increased to a gale, the sea beating heavily into the bay.' A fearful sea in Whitby Roads.

'Daily Telegraph' of 11th January—
>Jan. 10th, weather at Paris changed to a thaw.

At Chepstow (J. G. Wood, Esq.'s register)—
>Jan. 9th, thaw began about 9th Jan. at 9 A.M.
>„ 10th, thawing all day; cloudy, wind SE. to SW.
>„ 11th, thick small rain (characteristic).
>„ 12th, *very high tide* (as predicted) this morning.

1864.
Jan. 13th.
> Dead calm. The thick wet fog set in which is so characteristic of most lunar periods in the absence of strong wind. Great change of weather.
>
> At Chepstow, very thick fog all day, and dead calm.

„ 20th.
> Very strong SW. gale, worst at night, and continued with intermissions till noon of 23rd. Rapid fall of barometer.
>
> At Chepstow, damp, *foggy* forenoon; rain at mid-day. Mr. Wood noticed, on the previous night, 'disturbance in upper air.'
>
> At St. Leonards-on-Sea, Mr. T. Brett also noticed a perfect roar on night of 19th, with a *drizzly* fog on the 20th.

„ 27th.
> Fresh wind, W.; increased next day to a gale from NW.
>
> At Chepstow, squalls; high winds in afternoon; wind changed from S. to W.

Feb. 2nd.
> Fresh wind, SW. Exceedingly rapid scud (denoting a gale not far distant)—barometer falling—succeeded by heavy squalls of wind and much rain on the 3rd, with a tremendously hard and *sudden* squall of wind and rain at 8 P.M. of 3rd.
>
> At Unst, in the Shetland Isles, a 'terrible gale' on the 2nd, so furious as to blow a small boat, containing some hundreds weight of stone ballast, over another boat and on to a cow-shed, taking (so my son tells me) ten men to get it down again.

„ 7th to 9th (a period specially warned against by me in June)
> At Sheerness, a very quiet period, with *fall of snow on the 7th*; but from threats in the heavens and the barometer, Admiral Fitz Roy warned the coasts on the 8th, 9th, and 10th. The barometer continued

low and *uneasy*, with variable winds, until the very heavy gale of the 13th, which proved so destructive to shipping.

Tide not particularly high at Sheerness (as expected it would be, but I have not yet heard from the coast).

At Unst, Shetland Isles, the tide was very high.

1864.
February 16th.

Wind shifted from SSW. to NW., and frost set in, with tremendously heavy squalls, and rain; and the weather continued to be squally, with heavy snow storms, until the 22nd, when it fell dead calm, and the NW. to NE. wind subsided.

„ 23rd & 24th.

After some days' fine weather, a strong wind sprung up on 24th, lasting till 26th, with cold cloudy weather.

Mar. 2nd.

Remarkably dense wet fog, succeeded by heavy rains, and variable winds on ' second day after.'

„ 7th to 10th.

Proved a *very dangerous period, as specially warned against*; a complete hurricane in the English Channel, with very high tides.

„ 14th.

Very strong wind all night with permanent change of wind on 2nd day after, from Westerly to Easterly.

„ 22nd.

Strong gale, highest at 2 A.M. of 23rd, with variable SW. and NW. winds until the NE. wind of previous week *returned* on the 2nd day after.

I need not attempt to influence the reader's decision. As I said in the first instance, my object is *truth*, and a great public benefit, and *not self-interest*; and therefore, having given such ample details as to the groundwork of my assumed theory, if 'philosophers' can explain the 'lucky hits' and 'coincidences' by any other reasonable and palpable laws, I should not feel myself at all degraded by publicly confessing my error.

Now, those who have watched my writings on weather warning, will know that I consider all periods of probable atmospheric disturbance to be dangerous to shipping; but some are so in an increased degree. I explained this in the 'Nautical Magazine,' for so far back as January 1860 (vide page 22 therein): I said, 'such changes most commonly are accompanied either by strong winds, gales, *sudden frost, sudden thaw, sudden calms,* or other *certain interruptions of the weather*, according to the season.

While exceedingly thankful for a friendly suggestion that public defence of a theory against distinguished men might, in the eyes of the authorities, 'injure me,' I simply reply that from my *experience* of their sense of honour, I do not fear; I should not, however, be deterred from upholding my reputation in the *firm maintenance of right to an important discovery of which posterity will speak,* and this from two powerful motives—first, from its immense bearing on maritime commerce, and upon the increased saving of human life which must ensue; and next, because I am the parent of sons themselves as well-trained in honourable principles as in professional science, and I owe it to *their* future to vindicate my consistency and integrity.

CHAPTER XI.

SAXBY'S WEATHER SYSTEM EXPLAINED — CYCLONES AND THEIR CONNECTION WITH LUNAR INFLUENCES — REMARKABLE CYCLONE OF NOVEMBER 23, 1863, ILLUSTRATED AND DESCRIBED.

The following is what Europe now hears of as 'Saxby's Weather System: '—

I found that the moon never crosses the earth's equator or reaches her position of stitial colure, without a marked disturbance of the atmosphere occurring at the same period. Therefore I began by combining these lunar changes and actual weather disturbances into the relationship of '*cause and effect.*'

In considering the cause, there was little difficulty in tracing it to disturbance of the electric system of the globe at such periods, in consequence of changes in the effect produced in our atmosphere by variations in the intensity of the attraction of the sun or moon, or both, as they vary their positions with respect to the earth. I had before me as a guide the well-known and recognised influence of the moon's attraction upon the tides.

I find that the phases of the moon, properly so called, do not, perceptibly, affect the weather.

That the period of new moon has a marked influence on the weather when occurring near the periods of lunar equinox or lunar stitial colure, and that this influence is heightened considerably if at the same time the moon be in perigee. That the mere fact of her being new or in perigee is of no importance at any other times. So that *when the lunar equinox or stitial colure, occurs at the same period as the new moon in perigee, the greatest atmospheric disturbances to which our earth is liable may then be expected with certainty of fulfilment.*

Now, either the moon crosses our equator, or is at her stitial colure (or greatest distance from our equator) about once in a week (it is possible that one of each may happen in the same week, according to the inclination of her orbit), therefore there is an atmospheric disturbance from this cause, say once a week, and it is *invariably* traceable, as will presently be further explained.

So that this 'Weather System' is extremely simple. I fear many distrust it on account of such simplicity. For if we condense the definition of the above theory, it is mainly a mere question of the moon's declination.

As shown by reports from abroad, I believe that its effects are universal, even at the regions of the trade winds, but there with less power, and this is easy to be accounted for if we consider the cause of the air currents of the earth.

The method in which the weather is affected by the forces of the sun's and moon's attraction, is in the necessarily resulting sudden condensations of moisture in the atmosphere, or in the increased evaporation of moisture, according as the air is in a state of positive or negative electricity in a district, when disturbed by the sudden affection of the whole electric system at the periods above denoted.

Now, sudden condensation of moisture in the atmosphere produces a partial vacuum, into which surrounding air necessarily rushes; hence the periods of lunar equinox, or colures, are periods of 'atmospheric disturbance,' the intensity of such depending much on localities; and where such condensations or evaporations occur upon a large scale, some days may elapse before equilibrium is restored.

High winds at these periods are not always accompanied by falls of mercury in the barometer, since no accumulation of the electric fluid (which is imponderable) can increase the weight of the atmosphere.

A rising or steady barometer is not to be trusted at these periods, as severe gales frequently occur at such times without influencing that instrument. But a falling barometer at these times should never be disregarded.

Atmospheric disturbance manifests itself in ways not always recognised by an unscientific observer.

Where no absolute gale is blowing, its proximity is to be detected by the swift motion of 'scud,' (light, detached, fleecy clouds,) and, in the absence of these, there may often, at the lunar periods, be seen dark angry clouds packed in long thin horizontal lines (strata).

The months of April, May, June, and July show the lunar effects much more feebly than other months, although, as I have said, even in these, disturbance at lunar periods is plainly traceable by the meteorologist.

The above disturbances operating as they do *at the period of the lunar equinox, or stitial colure, being the immediate result of electric action,* I call *primary* or *electric*, to distinguish them from another kind of atmospheric disturbance, which I call *secondary* or *mechanical*, and for the following reasons, viz. :—

Whether from the nearer proximity of the equatorial regions of the earth to the moon, or from the peculiar action of what is termed 'thermo-electricity,' certain it is that such are the places which are subject to the greatest amount of precipitations or condensations. We may, therefore, easily conceive that atmospheric disturbances are there generated on the grandest scale. In those parts of the globe where it is known that the air contains the greatest amount of vapour (from the greater evaporations within the tropics), a sudden disturbance of electricity causing such an extensive precipitation, would require an immense flow of air to restore equilibrium, and that the air assumes a whirling motion is only what we see in all fluids under similar circumstances, but that this whirling should *always*, in the northern hemisphere, be in a direction contrary to the hands of a watch, is no more than we should expect from the circumstances of the earth's motion on its axis being from west to east, while on the southern side of the equator, the whirl is in a contrary direction. These whirlings of air, moreover, have a progressive motion, always in a direction from the equator.

Hence arises a means of distinguishing my theory into two distinct heads thus :—

The *primary* effects of disturbance are immediate and stationary.

The *secondary* are progressive, and they travel a considerable distance.

The latter are known by the name of Hurricanes, or Cyclones as they are also called, and they require some further illustration.

All seem to agree that Cyclones, which occasionally prove so destructive in the West Indies, come from the south-eastward, between the NE. trade winds and the equator, originating in the region of about 10° to 15° north latitude, and (it is worthy of remark) they seem to form in all cases upon the grand scale, at or *near the line of no magnetic variation, and commence their course upon it for some distance.* It is so not only with the Atlantic Cyclones, but (as I showed some years since in the 'Nautical Magazine,' and also before a section of the British Association at Cambridge, Professor Challis presiding,) it is also peculiar to those which derive their origin at or near the parallel of 10° south latitude, to the southward of Java. In corroboration of this I give the words of Admiral Fitz Roy, at page 65 in the 'Weather Book for 1863:'

'The hurricane, or Cyclone, is impelled to the west in *low* latitudes, because the tendency of both currents there is to the westward along the surface; although one, the tropical, is *much less so*, and becomes actually easterly near the tropic, after which its *equatorial* centrifugal force is more and more evident, while the westwardly tendency of the polar circuit diminishes, and therefore at that latitude hurricane Cyclones cease to move westward (re-curve), then go towards the pole, and subsequently almost easterly (in some cases), though commonly towards the northeastward, till they expand, disperse, or ascend. *It is a mistake to suppose* they travel very far,' &c.

And further (at page 112) he says:—

'Whatever may have been the duration of any one Cyclonic storm in the Atlantic, in the West Indies, or in the Indian Ocean, no instance has yet been obtained *here* of a definite and reliable character, of a rotatory gale lasting or *travelling beyond four days*.'

Now this opinion is contrary to that of Professor Dové, who declares that in general the velocity of progression of the centre increases ' as soon as the storm changes its course at the edge of the torrid zone,' to reach which usually occupies *nearly* the whole of that period of four days. Dové even gives instances of what *he believes* to be authenticated Cyclone tracks, in which some have occupied ten or eleven days in sweeping the West India Islands and reaching just southward of Newfoundland (e.g. August 17th, 1827). He even gives details (in his 'Law of Storms,' page 178) as under, and it is important to me to quote this, because my 'Weather System' seems to require a longer duration of Cyclones than four days, to which Admiral Fitz Roy would limit them. Professor Dové says:—

'The storm which commenced in the neighbourhood of Martinique, on August 17th, 1827, reached St. Martin and St. Thomas on the 18th, passed to the NE. of Hayti, on the 19th reached

Turk Islands, on the 20th the Bahamas, on the 21st and 22nd the coast of Florida and St. Carolina, on the 23rd and 24th Cape Hatteras, on the 25th Delaware, on the 26th Nantucket, on the 27th Sable Island, and Porpoise Bank on the 28th, and *thus travelled* 3,000 *nautical miles in eleven days.*'

But, further, the main branch of the equatorial current sets from the equator to the north-westward, and merges into the Florida stream, which takes the easterly direction in its full sweep across the Atlantic; and then, whether it be caused by the higher temperature of the Gulf Stream or by some magnetic or electric repulsion of the continent, the course of Cyclones seems to be absorbed into the line of direction of the Florida Gulf Stream, which shunts them as it were across several contiguous lines of equal magnetic variation, and places them upon others in the direction of which (or very nearly) they reach the British Islands. Thus Cyclones always approach us from the southward and westward, and proceed towards the north-eastward; and this known fact affords the Board of Trade the opportunity of warning the northern ports of an approaching Cyclone; for there is another peculiarity about Cyclones which ordinary gales are free from, and it can easily be explained. If we stir water in a tumbler, it will be seen that the centre of the fluid is depressed in proportion to the velocity of the fluid, and it is just so in the case of a Cyclone. The rapid whirling motion of the air yields to the law of centrifugal force and the air leaves the centre, consequently, within the vortex the height of the air becomes lowered, or we may say the *weight* of the air becomes lessened, so that it causes the mercury in a barometer to fall in consequence of diminished atmospheric pressure. And, further, the same law which induces a vessel in sailing through the water to push a small wave before it, causes the mass of whirling air, in its progress through the atmosphere, to push its atmospheric wave before it, which necessarily, by increasing the height of the column of air, slightly raises the mercury: this I have frequently noticed, and have called it the 'premonitory wave' of a Cyclone. I observed it at daylight on November 13th, 1861, and have often since, and early in the forenoon of that day *announced it in person to the underwriters of Lloyd's, as an indisputable proof that the Cyclone which I had predicted for the* 14*th was then not far from us*; and, to my knowledge, telegrams by London shipowners were sent to the coast to warn accordingly, and in time too, for, as before observed, the hurricane or Cyclone reached the Downs in its fury early on the 14th.

My reasons for so earnestly warning against the 14th were these: Cyclones, like all other general atmospheric disturbances, are *produced at the periods of lunar equinoxes and stitial colures*; and forasmuch as the stitial colure happened on the 5th, near the time of the *new moon in perigee* (before explained), I expected unusually well-developed Cyclones would be then generated in the regions of equinoctial calms; and again, because they travel at the rate of about sixteen to twenty miles an hour, and through a space of upwards of 4,000 miles before they reach us, I have found it very accurate to allow upon an average about nine days for the time of their passage to Great Britain; and as nine added to November 5 gave the 14th, I in consequence, as on other occasions, predicted accordingly. It frequently, however, happens that their passage occupies ten, eleven, or twelve days or more, for they travel at various velocities, so that when 'due' we should watch the barometer carefully, as we would for the arrival of a mail.

The surprise one naturally feels at the prolonged maintenance of the gyration of a Cyclone, diminishes when we notice that, according to Maury and others, as stated by Captain de Kerhallet (in 'Considerations Generales sur l'Ocean Atlantique,' and published in the 'Annales Hydrographiques' of 1852, and also in the 'Nautical' for 1855, &c.), no less than three branches of the main Atlantic Gulf Stream are directed entirely from it between the parallels of 40° and 50° north, thus, it is fair to assume, raising the temperature of the south-east side of the main current above that of the north-eastern side. Hence, then, if Cyclones travel along the main course of the gulf stream of the Atlantic, they will pass between *two masses of air of unequal density*; and inasmuch as the mass on the left hand or northern side is the denser, so will the whirl of the Cyclone be in a measure *maintained* towards the left hand, precisely in the direction it assumed at the time of its formation at the tropic. In other words, the Cyclone will *roll along* its boundary of denser air, as a ball would along a vertical plank which confined its lateral direction.

Now it appears also from Captain de Kerhallet that this main branch of the Gulf Stream crosses a point in the Atlantic at about 50° north latitude and 30° west longitude; but it would be highly interesting to see proof of the occurrence of Cyclones actually in that locality. For if we can only ascertain *beyond doubt* that Cyclones are ever found near that portion of the Atlantic, we may safely conclude that, inasmuch as all agree that Cyclones are formed in the Tropics—and the Admiral even explains the *mode of their formation there*—we may, I say, *safely assume that Cyclones found in lat. 50° north and long. 30° west have travelled from the tropics.* Of course, Cyclones in that district of 50° north and of 30° west would pass northward of Great Britain, and such moreover would be expected to prove heavy and dangerous to shipping.

It so happens that I can set this at rest if it be longer questioned. With that anxiety to help me (in the solution of so grand a problem as weather predicting) which has been singularly manifested by officers of the great mail fleets of the kingdom — and especially at Southampton, (whose interests in causing a theory so pretending as mine to be fairly tested no one can doubt — shared also by the members of the Dock Company), a letter, dated Dec. 18th, 1863, duly reached me with some particulars which I think of very great importance. I have not the pleasure of knowing Mr. Hedger, but he has certainly much assisted me in the confirmation of my opinion (and I thank him), by communicating an account from Captain Nicholson, of the Royal Mail steam-ship 'Adriatic,' on her homeward voyage in November last, as to a terrible Cyclone in which he was caught for seventeen hours, close to the very spot to which I have called attention, viz. at lat. 50° north and long. 30° west: of course this Cyclone was *on* the track of the said Gulf Stream, and precisely what we wanted to hear about.

The following are the particulars as furnished by Mr. Hedger:—

1863.

'Nov. 23rd. Noon. — Falling barometer — Wind variable and light from E. with drizzling rain.

,,   8 P.M. — A gale at ENE.

,,   12 P.M. — A heavy gale, NE.

,,   24th. 6 A.M. — Fearful hurricane and mountainous sea, wind NNE. Barometer 28·00 for 15 consecutive hours.

,,   8 P.M. — Wind NW. Glass rising.

,,   Midnight. — Squalls, less heavy.

,,   25th. A.M. — Moderating.

Now, landsmen have little idea of what we can make of such short notes, I would strongly advise those interested in the subject to purchase a little book* called

---

* Potter, Poultry. One Shilling.

the 'Storm Compass,' written by a masterly hand on this as on every other nautical subject, viz. by Captain A. B. Becher, R.N., F.R.A.S.

If we project this storm we shall find as under:—

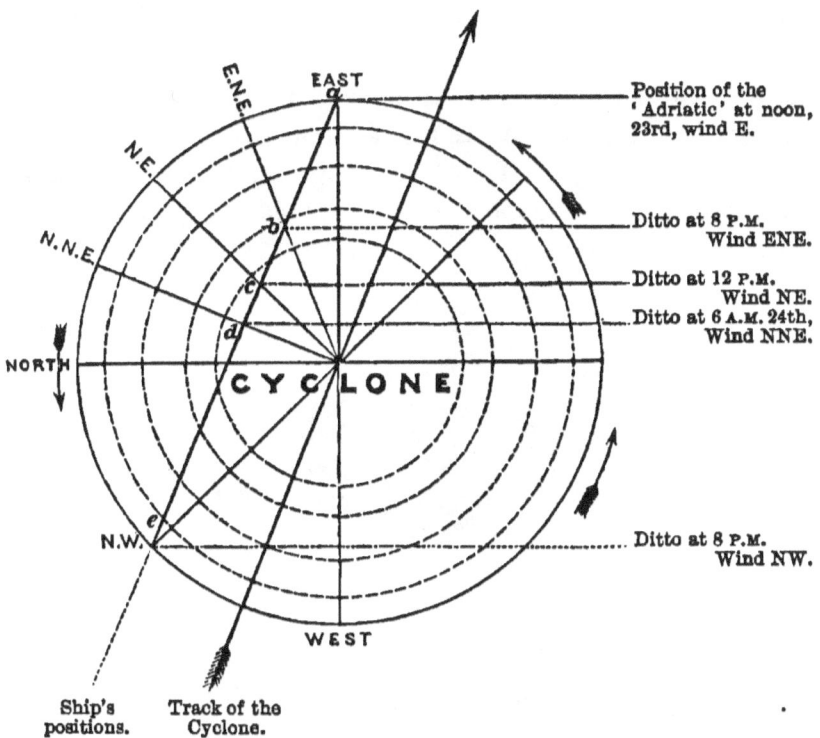

—That the 'Adriatic's' nearest proximity to the centre of the Cyclone was when she had the 'fearful hurricane' at NNE. (as at d in the accompanying figure), the centre then bearing ESE.

—That the Cyclone was apparently travelling about N 23° E. And it will be seen that the spaces $a\ b$, $b\ c$, &c. correspond to the intervals of time between the shiftings of the wind.

The outer circle being taken as the assumed limit of the Cyclone; the above is drawn as if the ship had been stationary the whole time she was in the Cyclone, but she was not. A little consideration will show that the actual course of the hurricane was more easterly.

Very erroneous notions are extant as to the form of these Cyclones; many suppose that viewed (if possible) from a distance they would appear like tall whirling cylinders or columns of air; but if we consider the diameter of some of them to be, say 500 miles, while their height can scarcely be more than a mile or two, such a Cyclone is merely a *disc* of rotating air, an *eddy* upon an enormous scale, deeper or thinner as indicated by the barometer, something like the accompanying figure.

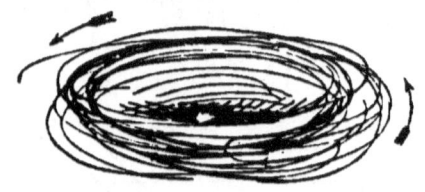

CHAPTER XII.

FULFILMENTS IN 1862 AND 1863—GREAT DAY AURORAL STORM OF JANUARY 1863—WEATHER WARNINGS UP TO JANUARY 1ST, 1866.

In conclusion, wishing to set forth as many facts only as may be necessary to illustrate the true state of the question of forewarning as carried out by me, I add the following from the 'Nautical,' of February 1863.

'I promised in my last letter to honestly acknowledge in your pages if it should be found that any one of my " predictions " had been unsupported by *fulfilment*; and specially mentioned the then approaching period of November 21st and 22nd, against which (among others) I had extensively warned in the August preceding.

'Probabilities were in favour of storm on November 22nd; but the great atmospheric disturbance of that period was felt rather in the North Atlantic Ocean than in Europe. These changes are upon a great scale of magnitude, and affect whole continents at a time. Our proofs of such disturbance were quite as satisfactory as, and infinitely more agreeable than, if the burden of the storm had passed over us. We in England and Ireland were only on the outskirts of it; but it passed westward or raged westward of us, and this is stated upon twofold authority. First, the barometer at Sheerness fell *at such period precisely* from 30·20 to 29·72 in.; while at the west coast of Ireland the barometer at the same time went to as low as 29·52 in. The weather at Sheerness undergoing the characteristic fall of temperature which I mention in all my printed lists. The scud also flew rapidly overhead, while on the earth's surface the weather was still and summerlike. At the same period further north, viz. at Unst, in

Shetland, an intense frost set in on this very same "lunar day." Secondly, the newspapers informed us that a heavy gale was blowing not far west of Scilly.

'If necessary I could add more, but such is enough to convince even the unwilling; perhaps, however, I may as well add, that the fineness of the day of lunar colure was succeeded on the "second day after," according to what I always expect and for three years have called attention to, viz. on November 24th, by a gale. A more conclusive fulfilment could not be desired. *I beg to distinguish between positive changes and positive gales.*

'The next marked day in my list was November 29th, only marked as of ordinary disturbance; yet even this was an equally well attested period of change. There was the usual *uneasy* changing, shifting of the wind and the well marked depression of the barometer, the wind during the day flying all round the compass.

'On December 6th (the next lunar day) we had the similar change of wind, &c., with fresh gale and rain.

'13th and 14th (next lunar days).—Heavy gale in the Channel, with driving rain. Barometer fell ·46 in. Fine aurora borealis at 9 P.M.

'20th and 21st (next lunar days).—Terrific storms, with thunder, hail, &c. At 3 P.M. of 21st velocity of wind seventy miles per hour. The greatest depression of barometer during the gale was ·97 in. Extraordinary high tide on 21st. (*I had for weeks beforehand warned publicly both against storm and high tide on these very days.*)

'26th (next lunar day).—Gale and characteristic barometric depression.

'On January 1st, 1863, the "Standard" newspaper, published days of expected change up to July 1st next. Of those for January and February have already occurred the following fulfilments at Sheerness:—

'January 2nd.—Heavy gale, with much rain.

'10th.—Moderate gale. Fall of barometer was ·25 in. Lower temperature set in.

'18th and 29th.—Most destructive and terrible gale, with *very high tide* on the 20th everywhere. From noon to 7 P.M. of the 20th the velocity of the wind was above seventy miles an hour. Between noon of 17th and 9 A.M. of the 18th the barometer fell ·86 in.; and from 4.30 of 19th to 6 A.M. of the 20th it fell ·55 in., the lowest point of barometer during the gale having been 28·97 in. (corrected for temperature and sea level). An extraordinary "luminosity" at 10 A.M. of 20th in the NNW. part of the heavens.

'(N.B. So strongly did I expect this serious weather that during *weeks beforehand* I distributed five hundred special printed warnings as to 18th, 19th, and 20th, particularly cautioning against the high tide of the 20th. I also warned through several local papers.)

'30th.—Very heavy gale. Velocity of wind at least sixty miles per hour. Fall of barometer between 11h. 30m. of 29th and 1h. P.M. of 31st was ·51 in.

'February 6th.—Disturbance aloft, indicated by very rapid scud from westward, while on the earth's surface it was a perfect summer's day. From 11 P.M. it blew hard outside the Thames. Two "parhelia" visible in the afternoon (see "Standard" of February 9th). Barometer attained its maximum height at 3 P.M., it being 30·34.

'N.B. At Unst "deluge of rain; almost a hurricane from West."

'13th.—Wind all round the compass on the 12th. Very strong scud from westward on 13th, although little wind at Sheerness and fine weather. During the day the wind changed from NE., at 9 A.M., to E. b. S., at 10.30 A.M., which continued several days.

'18th and 19th.—On 18th, after four cloudless days and calm weather, clouds began to form at early A.M.; by 3 P.M. cloudy. Wind on 17th and 18th very variable; a most gorgeous sunset on the 18th.

'On 19th (the actual day of lunar equinox) the change of weather was remarkable. Instead of the fine weather which set in on the 13th (the previous lunar day) the weather on the 19th *totally changed*, with a light wind from NW. and a high, slightly falling, barometer. Heavy rain set in at 4 A.M., with thick haze and continued drizzle, and a dead calm lasted all day. Nothing could be more marked). Sunshine returned on the 20th; barometer singularly high and steady.

'The above details of each occurring "lunar period," &c., are given as data for comparisons to those who are closely watching my theory in different parts of England. I need not ask if my absolute "predictions" were justifiable, their fulfilment through the whole period having *been strictly and literally indisputable.* I will not condescend to call them "forecasts," since they are founded on a law which, though newly discovered and only as yet imperfectly developed, is now established and ought to be unassailable.

'As regards the unusually high tides of this winter, they are the natural consequence of the moon and sun both being in perigee as connected with declination, the former being also at its position of new moon. In December last such happened; moreover, almost precisely at the time of lunar stitial colure (within a few hours): hence the greater rise of the high December tide (irrespective of winds and local direction); and also the diminished rise in January, when the new moon in perigee fell between the corresponding colure and the lunar equinox, although the atmospheric disturbance seems to have been nearly equal in each case.

'I specially cautioned the coast, both by letters and precise advertisement, against the storms and high tides of December and January, but not against the corresponding period in the present month, because, in the first place, the sun is now, in February, rapidly receding (so to speak) from the earth, and his disturbing power is therefore rapidly diminishing, inversely as the cube of his distance; and in the second place because the new moon of the day before the lunar equinox of the 19th would, on the day of such equinox, be some 50° of her orbit distant from her place of perigee. But still I considered it ought to be marked at least as a "black letter" day in my printed list of warnings.

'Very extraordinary were the gales between the 16th and 23rd January (as I widely forewarned). Whether we regard the intensity of the gales or their duration, one of them will henceforward rank with the most painfully prominent storms of this era. That of October 1859, will long be known as the "Royal Charter Gale," but this of 20th January last quite equals it in intensity and in interest, accompanied, as it was, by such extraordinary tides on all parts of the coast. Even in Shetland, where the rise and fall is trifling, it was felt in astonishment. When one of my sons delivered my warning to the fishermen there, it was of course received with respect; but when the tide rose, their amazement knew no bounds. In their simplicity their awe of him was ludicrous.

'But that which preeminently ought to render it memorable is the extraordinary "luminosity" already spoken of (and if I stretch my limit as to space in the "Nautical" to the very extreme, pray forgive me). At 10 A.M. of the 20th January, when the velocity of wind, measured by cup and dial anemometer, was nearly seventy miles an hour, I noticed on reaching the quarter-deck that a peculiar light (I think I once saw a somewhat similar light off Cape St. Mary, Madagascar) seemed to shine upon all the NW. quarter of the horizon. The town of Southend, on the Essex shore (not by any means a prominent object at Sheerness), was so plainly visible that its terraces of houses were startlingly conspicuous. Everything also on the nearer Isle of Grain seemed to "stand out" in the almost *spectral glow,* which prevailed at that part of the sky only. To heighten the contrast, *shreds* of highly electric racing fragments of fleecy or hairy clouds of a deep, dark colour arrested the attention from their singular filamentary density. These clouds could only, from their velocity, have been a hundred or two yards in altitude. Behind them one would have supposed the sun to have been momentarily "muffled," so *bright* rather than *light* was the appearance as compared with the gloom of dense masses of clouds in all other directions.

'Now, Sir, reference to the diagram will show that 10 A.M. was just before or at the very moment of the height of the fury of the storm (this height lasting for nine hours! for it was not till 7 P.M. that the wind abated).

'Much perplexed during the day at the singular luminosity referred to, and striving to reconcile it with any previous experiences, I recalled to mind again and again the fact of one part alone of the heavens being then so free from clouds, excepting the electric scud referred to. But during my musings in the evening of the same day, it occurred to me that it must have been an aurora borealis of such brilliancy as to have been plainly visible by daylight. Once satisfied as to this, there was little difficulty in tracing the connection between Sir John Herschel's "polar currents" and their occasional "downrushes," as quoted by Admiral Fitz Roy. If ever the eyes of man beheld such a "polar downrush," mine did on the 20th January last, and therefore I call that tremendous gale an "auroral storm."

'That the worst of a gale is generally felt an hour or two after the barometer has passed its greatest depression is an accepted rule among sailors. In this case it will be seen that the lowest barometer of the gale happened at 6 A.M. of the 20th, when it had fallen to 28·87 in. (accurately corrected). By 10 o'clock it had risen to 29·05 in.

'If your readers will kindly notice, the heavy gale of the 16th January occurred on one of my "lunar days;" and its distinctness in character is worthy of remark, for in it the wind was NE.; in the gale of the 18th, 19th, and 20th, and part of 21st, the wind throughout was WNW.; while in the succeeding gale of the 22nd (the next lunar day) the wind was from SW. Such distinct corroborations of a theory cannot be expected in every month; but the past three months have been of so peculiar a nature that all the energies of accessories to atmospheric disturbance have been noted under advantageous circumstances, and their attributes stored for future use. The season has, I have truly said, been a *trying one.*

'It would have been a triumph if a careful and elaborate collation of reports from a whole continent had resulted in the detection of a lunar weather system. But a still greater triumph is it for one unaided observer at a single station to have accomplished what I submit is now proved to be so great a discovery.

'Allow me to congratulate your readers (and myself too?) upon the great success I have achieved,—an advantage not merely to the present generation, but one which, when I am in my grave, will be acknowledged by posterity.

'I have, &c.

'S. M. SAXBY, R.N.'

YET UNFULFILLED PREDICTIONS.

Having thus, as I believe, vindicated and established my claim to the discovery of a 'Lunar Weather System,' and sufficiently explained its nature, I leave with the reader a list of suspected periods up to 1st January 1866.

List of Days on which the weather may reasonably be suspected as *liable to change, most probably towards high winds or lower temperature*, being especially periods of *atmospheric disturbance*.

1864.—March ..... 2nd—**7th to 10th**—14th—22nd—29th.
          N.B. The 7th and 8th will be a very dangerous period, with probably a very full tide on the 10th.
,,   April ...... **4th to 6th**—11th—18th—25th.
,,   May ...... **1st**—8th—15th—22nd—29th.
,,   June ...... **4th**—12th—19th—25th.
,,   July ...... 2nd—9th—16th—22nd—29th.
,,   August .... 5th—13th—19th—25th.
,,   September .. 2nd—9th—**15th**—22nd—29th.
,,   October .... 6th—**13th to 15th**—19th—26th.
,,   November ... 2nd and 3rd—9th—15th—23rd—30th.
,,   December ... 6th—13th—20th—27th.
1865.—January .... 3rd—9th—16th—24th—30th.
,,   February ... 5th—13th—20th—**26th**.
          N.B. February 26th is likely to be a very dangerous period, with probably a high tide.
,,   March ..... 5th—12th—19th—**26th**.
          N.B. March 26th will probably be a dangerous period, with high tide.
,,   April ...... 1st—8th—16th—22nd—28th.
          N.B. The 24th to 28th may very likely prove greatly unsettled.
,,   May .... 6th—13th—20th—26th.
          N.B. From 20th to 26th is likely to be a very unsettled period.
,,   June ...... 2nd—9th—16th—22nd—29th.
,,   July. ...... 7th—13th—20th—27th.
,,   August .... 3rd—9th—16th—23rd—30th.
,,   September .. **6th**—12th—19th—27th.
,,   October .... 3rd—9th—16th—24th—31st.
,,   November .. 6th—13th—20th—27th.
,,   December ... 3rd—10th—18th—24th—31st.

The preceding apply to all parts of the earth's surface—even (in a diminished degree) to the trade belts.

N.B. If the day marked prove calm and still, distrust the day after, and *especially the second day after*.

The changes vary in intensity, but even at *quiet* periods they may be plainly traced in the scud flying with a velocity totally at variance with the state of the air at the earth's surface, and the clouds at such times generally have a liny or stratified appearance, which usually indicates approaching rain.

N.B. As a general rule, electric agency is feeble in May, June, and July, as compared with other months. If, therefore, any changes take place in those months, they will most likely happen on some of the days marked.

My own impression is that the winters of 1864 and 1865 are more likely to prove frosty than windy.

In bringing the above to the test of experience, a few memoranda may prevent unintentional injustice towards the system.

1. Do not consider the prediction a failure if the weather prove moderate. The above days are dates of *change*, not of *necessity* periods of bad weather.

2. Watch for changes of wind at such times, and what sailors call its *uneasiness*—flying about from point to point with unsteadiness.

3. A characteristic of these lunar periods (when they prove quiet) is *fog* and *haze*, very often setting in the day before or a few hours before the date given.

4. When the mercury falls for a day or two before a lunar day, expect its effects to last just about so long afterwards. The lunar period being the middle of the time of disturbance.

5. The barometer *is not always affected* by those lunar periods —but there is a tendency in the mercury to *either change its direction* up or down at such times, or the apex of a curve projected by it will happen at such period.

6. The rising of the barometer on or after the 'second day after' generally indicates the returning of fine weather.

EXPLANATION OF TERMS.

7. If, on or about the 'second day after,' the mercury fall rapidly, prepare for a Cyclone. (This only applies to the Atlantic district, and the western coast of France, England, Ireland, &c., on reaching which their presence is generally announced by Admiral Fitz Roy's coast warnings.)

S. M. S.

The following illustrate the terms used in relation to the Moon's positions with respect to the Sun and Earth.

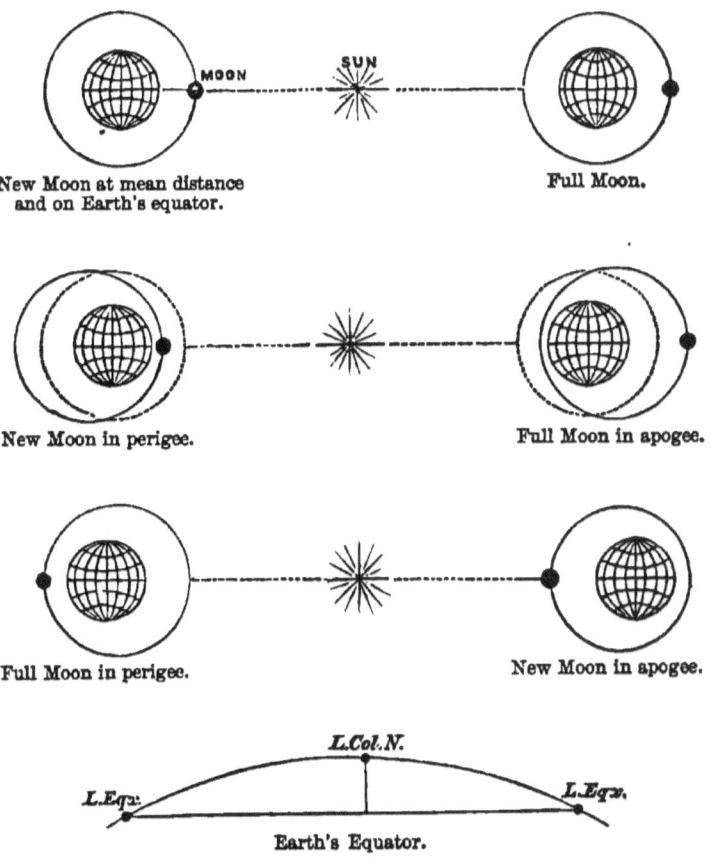

New Moon at mean distance and on Earth's equator.　　Full Moon.

New Moon in perigee.　　Full Moon in apogee.

Full Moon in perigee.　　New Moon in apogee.

L.Col.N.

L.Eqx.　　　　　　　　　　*L.Eqx.*

Earth's Equator.

PRINTED BY SPOTTISWOODE AND CO., NEW-STREET SQUARE, LONDON

[JULY 1864.]

GENERAL LIST OF WORKS

PUBLISHED BY

MESSRS. LONGMAN, GREEN, AND CO.

PATERNOSTER ROW, LONDON.

Historical Works.

The **HISTORY of ENGLAND** from the Fall of Wolsey to the Death of Elizabeth. By JAMES ANTHONY FROUDE, M.A. late Fellow of Exeter College, Oxford. Third Edition of the First Eight Volumes.

    VOLS. I to IV. the Reign of Henry VIII. Third Edition, 54s.

    VOLS. V. and VI. the Reigns of Edward VI. and Mary. Third Edition, 28s.

    VOLS. VII. and VIII. the Reign of Elizabeth, VOLS. I. and II. Third Edition, 28s.

The **HISTORY of ENGLAND** from the Accession of James II. By Lord MACAULAY. Three Editions as follows.

    LIBRARY EDITION, 5 vols. 8vo. £4.

    CABINET EDITION, 8 vols. post 8vo. 48s.

    PEOPLE'S EDITION, 4 vols. crown 8vo. 16s.

REVOLUTIONS in ENGLISH HISTORY. By ROBERT VAUGHAN, D.D. 3 vols. 8vo. 45s.

    VOL. I. Revolutions of Race, 15s.

    VOL II. Revolutions in Religion, 15s.

    VOL. III. Revolutions in Government, 15s.

The **HISTORY of ENGLAND** during the Reign of George the Third. By WILLIAM MASSEY, M.P. 4 vols. 8vo. 48s.

The **CONSTITUTIONAL HISTORY of ENGLAND**, since the Accession of George III. 1760—1860. By THOMAS ERSKINE MAY, C.B. 2 vols. 8vo. 33s.

LIVES of the QUEENS of ENGLAND, from State Papers and other Documentary Sources: comprising a Domestic History of England from the Conquest to the Death of Queen Anne. By AGNES STRICKLAND. Revised Edition, with many Portraits. 8 vols. post 8vo. 60s.

LECTURES on the HISTORY of ENGLAND. By WILLIAM LONGMAN. Vol. I. from the earliest times to the Death of King Edward II. with 6 Maps, a coloured Plate, and 53 Woodcuts. 8vo. 15s.

A CHRONICLE of ENGLAND, from B.C. 55 to A.D. 1485; written and illustrated by J. E. DOYLE. With 81 Designs engraved on Wood and printed in Colours by E. Evans. 4to. 42s.

HISTORY of CIVILISATION. By HENRY THOMAS BUCKLE. 2 vols. Price £1 17s.

Vol. I. *England and France,* Fourth Edition, 21s.

Vol. II. *Spain and Scotland,* Second Edition, 16s.

DEMOCRACY in AMERICA. By ALEXIS DE TOCQUEVILLE. Translated by HENRY REEVE, with an Introductory Notice by the Translator. 2 vols. 8vo. 21s.

The SPANISH CONQUEST in AMERICA, and its Relation to the History of Slavery and to the Government of Colonies. By ARTHUR HELPS. 4 vols. 8vo. £3. Vols. I. and II. 28s. Vols. III. and IV. 16s. each.

HISTORY of the REFORMATION in EUROPE in the Time of Calvin. By J. H. MERLE D'AUBIGNE, D.D. Vols. I. and II. 8vo. 28s. and Vol. III. 12s.

LIBRARY HISTORY of FRANCE, in 5 vols. 8vo. By EYRE EVANS CROWE. Vol. I. 14s. Vol. II. 15s. Vol. III. 18s. Vol. IV. nearly ready.

LECTURES on the HISTORY of FRANCE. By the late Sir JAMES STEPHEN, LL.D. 2 vols. 8vo. 24s.

The HISTORY of GREECE. By C. THIRLWALL, D.D., Lord Bishop of St. David's. 8 vols. 8vo. £3; or in 8 vols. fcp. 28s.

The TALE of the GREAT PERSIAN WAR, from the Histories of *Herodotus.* By the Rev. G. W. Cox, M.A. late Scholar of Trin. Coll. Oxon. Fcp. 8vo. 7s. 6d.

ANCIENT HISTORY of EGYPT, ASSYRIA, and BABYLONIA. By the Author of 'Amy Herbert.' Fcp. 8vo. 6s.

CRITICAL HISTORY of the LANGUAGE and LITERATURE of Ancient Greece. By WILLIAM MURE, of Caldwell. 5 vols. 8vo. £3 9s.

HISTORY of the LITERATURE of ANCIENT GREECE. By Professor K. O. MÜLLER. Translated by the Right Hon. Sir GEORGE CORNEWALL LEWIS, Bart. and by J.W. DONALDSON, D.D. 3 vols. 8vo. 36s.

HISTORY of the ROMANS under the EMPIRE. By the Rev. CHARLES MERIVALE, B.D. 7 vols. 8vo. with Maps, £5.

The FALL of the ROMAN REPUBLIC: a Short History of the Last Century of the Commonwealth. By the Rev. CHARLES MERIVALE, B.D. 12mo. 7s. 6d.

The BIOGRAPHICAL HISTORY of PHILOSOPHY, from its Origin in Greece to the Present Day. By GEORGE HENRY LEWES. Revised and enlarged Edition. 8vo. 16s.

HISTORY of the INDUCTIVE SCIENCES. By WILLIAM WHEWELL, D.D. F.R.S. Master of Trin. Coll. Cantab. Third Edition. 3 vols. crown 8vo. 24s.

CRITICAL and HISTORICAL ESSAYS contributed to the *Edinburgh Review*. By the Right Hon. LORD MACAULAY.
   LIBRARY EDITION, 3 vols. 8vo. 36s.
   TRAVELLER'S EDITION, in 1 vol. 21s.
   In POCKET VOLUMES, 3 vols. fcp. 21s.
   PEOPLE'S EDITION, 2 vols. crown 8vo. 8s.

EGYPT'S PLACE in UNIVERSAL HISTORY; an Historical Investigation. By C. C. J. BUNSEN, D.D. Translated by C. H. COTTRELL, M.A. With many Illustrations. 4 vols. 8vo. £5 8s. VOL. V. is nearly ready.

MAUNDER'S HISTORICAL TREASURY; comprising a General Introductory Outline of Universal History, and a series of Separate Histories. Fcp. 8vo. 10s.

HISTORICAL and CHRONOLOGICAL ENCYCLOPÆDIA, presenting in a brief and convenient form Chronological Notices of all the Great Events of Universal History. By B. B. WOODWARD, F.S.A. Librarian to the Queen.
                                                                              [*In the press.*

HISTORY of CHRISTIAN MISSIONS; their Agents and their Results By T. W. M. MARSHALL. 2 vols. 8vo. 24s.

HISTORY of the EARLY CHURCH, from the First Preaching of the Gospel to the Council of Nicæa, A.D. 325. By the Author of 'Amy Herbert.' Fcp. 8vo. 4s. 6d.

HISTORY of WESLEYAN METHODISM. By GEORGE SMITH, F.A.S. New Edition, with Portraits, in 31 parts. Price 6d. each.

HISTORY of MODERN MUSIC; a Course of Lectures delivered at the Royal Institution. By JOHN HULLAH, Professor of Vocal Music in King's College and in Queen's College, London. Post 8vo. 6s. 6d.

HISTORY of MEDICINE, from the Earliest Ages to the Present Time. By EDWARD MERYON, M.D. F.G.S. Vol. I. 8vo. 12s. 6d.

Biography and Memoirs.

SIR JOHN ELIOT, a Biography; 1590—1632. By JOHN FORSTER. With Two Portraits on Steel from the Originals at Port Eliot. 2 vols. crown 8vo. 30s.

LETTERS and LIFE of FRANCIS BACON, including all his Occasional Works. Collected and edited, with a Commentary, by J. SPEDDING, Trin. Coll. Cantab. VOLS. I. and II. 8vo. 24s.

LIFE of ROBERT STEPHENSON, F.R.S. By J. C. JEAFFRESON, Barrister-at-Law; and WILLIAM POLE, F.R.S. Memb. Inst. Civ. Eng. With 2 Portraits and many Illustrations. 2 vols. 8vo. [*Nearly ready.*

APOLOGIA pro VITA SUA: being a Reply to a Pamphlet entitled 'What then does Dr. Newman mean?' By JOHN HENRY NEWMAN, D.D. 8vo. 14s.

LIFE of the DUKE of WELLINGTON. By the Rev. G. R. Gleig, M.A. Popular Edition, carefully revised; with copious Additions. Crown 8vo. 5s.

Brialmont and Gleig's Life of the Duke of Wellington. 4 vols. 8vo. with Illustrations, £2 14s.

Life of the Duke of Wellington, partly from the French of M. Brialmont, partly from Original Documents. By the Rev. G. R. Gleig, M.A. 8vo. with Portrait, 15s.

FATHER MATHEW: a Biography. By John Francis Maguire, M.P. Second Edition, with Portrait. Post 8vo. 12s. 6d.

Rome; its Ruler and its Institutions. By the same Author. New Edition in preparation.

LIFE of AMELIA WILHELMINA SIEVEKING, from the German. Edited, with the Author's sanction, by Catherine Winkworth. Post 8vo. with Portrait, 12s.

FELIX MENDELSSOHN'S LETTERS from *Italy and Switzerland*, translated by Lady Wallace, Third Edition, with Notice of Mendelssohn's Life and Works, by Henry F. Chorley; and *Letters from 1833 to 1847*, translated by Lady Wallace. New Edition, with Portrait. 2 vols. crown 8vo. 5s. each.

DIARIES of a LADY of QUALITY, from 1797 to 1844. Edited, with Notes, by A. Hayward, Q.C. Second Edition. Post 8vo. 10s. 6d.

RECOLLECTIONS of the late WILLIAM WILBERFORCE, M.P. for the County of York during nearly 30 Years. By J. S. Harford, D.C.L. F.R.S. Post 8vo. 7s.

LIFE and CORRESPONDENCE of THEODORE PARKER. By John Weiss. With 2 Portraits and 19 Wood Engravings. 2 vols. 8vo. 30s.

SOUTHEY'S LIFE of WESLEY. Fifth Edition. Edited by the Rev. C. C. Southey, M.A. Crown 8vo. 7s. 6d.

THOMAS MOORE'S MEMOIRS, JOURNAL, and CORRESPONDENCE. Edited and abridged from the First Edition by Earl Russell. Square crown 8vo. with 8 Portraits, 12s. 6d.

MEMOIR of the Rev. SYDNEY SMITH. By his Daughter, Lady Holland. With a Selection from his Letters, edited by Mrs. Austin. 2 vols. 8vo. 28s.

LIFE of WILLIAM WARBURTON, D.D. Bishop of Gloucester from 1760 to 1779. By the Rev. J. S. Watson, M.A. 8vo. with Portrait, 18s.

FASTI EBORACENSES: Lives of the Archbishops of York. By the late Rev. W. H. Dixon, M.A. Edited and enlarged by the Rev. J. Raine, M.A. In 2 vols. Vol. I. comprising the lives to the Death of Edward III. 8vo. 15s.

VICISSITUDES of FAMILIES. By Sir Bernard Burke, Ulster King of Arms. First, Second, and Third Series. 3 vols. crown 8vo. 12s. 6d. each.

BIOGRAPHICAL SKETCHES. By NASSAU W. SENIOR. Post 8vo. price 10s. 6d.

ESSAYS in ECCLESIASTICAL BIOGRAPHY. By the Right Hon. Sir J. STEPHEN, LL.D. Fourth Edition. 8vo. 14s.

ARAGO'S BIOGRAPHIES of DISTINGUISHED SCIENTIFIC MEN. By FRANÇOIS ARAGO. Translated by Admiral W. H. SMYTH, F.R.S. the Rev. B. POWELL, M.A. and R. GRANT, M.A. 8vo. 18s.

MAUNDER'S BIOGRAPHICAL TREASURY: Memoirs, Sketches, and Brief Notices of above 12,000 Eminent Persons of All Ages and Nations. Fcp. 8vo. 10s.

Criticism, Philosophy, Polity, &c.

PAPINIAN: a Dialogue on State Affairs between a Constitutional Lawyer and a Country Gentleman about to enter Public Life. By GEORGE ATKINSON, B.A. Oxon. Serjeant-at-Law. Post 8vo. 5s.

On REPRESENTATIVE GOVERNMENT. By JOHN STUART MILL. Second Edition, 8vo. 9s.

Dissertations and Discussions. By the same Author. 2 vols. 8vo. price 24s.

On Liberty. By the same Author. Third Edition. Post 8vo. 7s. 6d.

Principles of Political Economy. By the same. Fifth Edition. 2 vols. 8vo. 30s.

A System of Logic, Ratiocinative and Inductive. By the same. Fifth Edition. Two vols. 8vo. 25s.

Utilitarianism. By the same. 8vo. 5s.

LORD BACON'S WORKS, collected and edited by R. L. ELLIS, M.A. J. SPEDDING, M.A. and D. D. HEATH. Vols. I. to V. *Philosophical Works* 5 vols. 8vo. £4 6s. VOLS. VI. and VII. *Literary and Professional Works* 2 vols. £1 16s.

BACON'S ESSAYS with ANNOTATIONS. By R. WHATELY, D.D. late Archbishop of Dublin. Sixth Edition. 8vo. 10s. 6d.

ELEMENTS of LOGIC. By R. WHATELY, D.D. late Archbishop of Dublin. Ninth Edition. 8vo. 10s. 6d. crown 8vo. 4s. 6d.

Elements of Rhetoric. By the same Author. Seventh Edition. 8vo. 10s. 6d. crown 8vo. 4s. 6d.

English Synonymes. Edited by Archbishop WHATELY. 5th Edition. Fcp. 8vo. 3s.

MISCELLANEOUS REMAINS from the Common-place Book of the late Archbishop WHATELY. Edited by Miss E. J. WHATELY. Post 8vo. 6s.

ESSAYS on the ADMINISTRATIONS of GREAT BRITAIN from 1783 to 1830, contributed to the *Edinburgh Review* by the Right Hon. Sir G. C. LEWIS, Bart. Edited by the Right Hon. Sir E. HEAD, Bart. 8vo. with Portrait, 15s.

By the same Author.

A Dialogue on the Best Form of Government, 4s. 6d.

Essay on the Origin and Formation of the Romance Languages, price 7s. 6d.

Historical Survey of the Astronomy of the Ancients, 15s.

Inquiry into the Credibility of the Early Roman History, 2 vols. price 30s.

On the Methods of Observation and Reasoning in Politics, 2 vols. price 28s.

Irish Disturbances and Irish Church Question, 12s.

Remarks on the Use and Abuse of some Political Terms, 9s.

On Foreign Jurisdiction and Extradition of Criminals, 2s. 6d.

The Fables of Babrius, Greek Text with Latin Notes, PART I. 5s. 6d. PART II. 3s. 6d.

Suggestions for the Application of the Egyptological Method to Modern History, 1s.

An OUTLINE of the NECESSARY LAWS of THOUGHT: a Treatise on Pure and Applied Logic. By the Most Rev. W. THOMSON, D.D. Archbishop of York. Crown 8vo. 5s. 6d.

The ELEMENTS of LOGIC. By THOMAS SHEDDEN, M.A. of St. Peter's Coll. Cantab. Crown 8vo. [*Just ready.*

ANALYSIS of Mr. MILL'S SYSTEM of LOGIC. By W. STEBBING, M.A. Fellow of Worcester College, Oxford. Post 8vo. [*Just ready.*

SPEECHES of the RIGHT HON. LORD MACAULAY, corrected by Himself. 8vo. 12s.

LORD MACAULAY'S SPEECHES on PARLIAMENTARY REFORM in 1831 and 1832. 16mo. 1s.

A DICTIONARY of the ENGLISH LANGUAGE. By R. G. LATHAM, M.A. M.D. F.R.S. Founded on that of Dr. JOHNSON, as edited by the Rev. H. J. TODD, with numerous Emendations and Additions. Publishing in 36 Parts, price 3s. 6d. each, to form 2 vols. 4to.

The English Language. By the same Author. Fifth Edition. 8vo. price 18s.

Handbook of the English Language. By the same Author. Fourth Edition. Crown 8vo. 7s. 6d.

Elements of Comparative Philology. By the same Author. 8vo. 21s.

THESAURUS of ENGLISH WORDS and PHRASES, classified and arranged so as to facilitate the Expression of Ideas, and assist in Literary Composition. By P. M. ROGET, M. D. 14th Edition. Crown 8vo. 10s. 6d.

LECTURES on the SCIENCE of LANGUAGE, delivered at the Royal Institution. By MAX MULLER, M.A. Fellow of All Souls College, Oxford. FIRST SERIES, Fourth Edition. 8vo. 12s. SECOND SERIES, with 31 Woodcuts, price 18s.

The **DEBATER**; a Series of Complete Debates, Outlines of Debates, and Questions for Discussion. By F. ROWTON. Fcp. 8vo. 6s.

A **COURSE of ENGLISH READING**, adapted to every taste and capacity; or, How and What to Read. By the Rev. J. PYCROFT, B.A. Fcp. 8vo. 5s.

MANUAL of ENGLISH LITERATURE, Historical and Critical: with a Chapter on English Metres. By T. ARNOLD, B.A. Prof. of Eng. Lit. Cath. Univ. Ireland. Post 8vo. 10s. 6d.

SOUTHEY'S DOCTOR, complete in One Volume. Edited by the Rev. J. W. WARTER, B.D. Square crown 8vo. 12s. 6d.

HISTORICAL and CRITICAL COMMENTARY on the OLD TESTAMENT; with a New Translation. By M. M. KALISCH, Ph.D. VOL. I. *Genesis*, 8vo. 18s. or adapted for the General Reader, 12s. VOL. II. *Exodus*, 15s. or adapted for the General Reader, 12s.

A Hebrew Grammar, with Exercises. By the same. PART I. *Outlines with Exercises*, 8vo. 12s. 6d. KEY, 5s. PART II. *Exceptional Forms and Constructions*, 12s. 6d.

A NEW LATIN-ENGLISH DICTIONARY. By the Rev. J. T. WHITE, M.A. of Corpus Christi College, and Rev. J. E. RIDDLE, M.A. of St. Edmund Hall, Oxford. Imperial 8vo. 42s.

A Diamond Latin-English Dictionary, or Guide to the Meaning, Quality, and Accentuation of Latin Classical Words. By the Rev. J. E. RIDDLE, M.A. 32mo. 4s.

A NEW ENGLISH-GREEK LEXICON, containing all the Greek Words used by Writers of good authority. By C. D. YONGE, B.A. Fourth Edition. 4to. 21s.

A LEXICON, ENGLISH and GREEK, abridged for the Use of Schools from his 'English-Greek Lexicon' by the Author, C. D. YONGE, B.A. Square 12mo. [*Just ready.*

A GREEK-ENGLISH LEXICON. Compiled by H. G. LIDDELL, D.D. Dean of Christ Church, and R. SCOTT, D.D. Master of Balliol. Fifth Edition. Crown 4to. 31s. 6d.

A Lexicon, Greek and English, abridged from LIDDELL and SCOTT's *Greek-English Lexicon*. Tenth Edition. Square 12mo. 7s. 6d.

A PRACTICAL DICTIONARY of the FRENCH and ENGLISH LANGUAGES. By L. CONTANSEAU. 7th Edition. Post 8vo. 10s. 6d.

Contanseau's Pocket Dictionary, French and English; being a close Abridgment of the above, by the same Author. 2nd Edition. 18mo. 5s.

NEW PRACTICAL DICTIONARY of the GERMAN LANGUAGE; German-English and English-German. By the Rev. W. L. BLACKLEY, M.A. and Dr. CARL MARTIN FRIEDLANDER. Post 8vo. [*In the press.*

Miscellaneous Works and Popular Metaphysics.

RECREATIONS of a COUNTRY PARSON: being a Selection of the Contributions of A. K. H. B. to *Fraser's Magazine*. SECOND SERIES. Crown 8vo. 3s. 6d.

The Common-place Philosopher in Town and Country. By the same Author. Crown 8vo. 3s. 6d.

Leisure Hours in Town; Essays Consolatory, Æsthetical, Moral, Social, and Domestic. By the same. Crown 8vo. 3s. 6d.

The Autumn Holidays of a Country Parson. By the same Author. 1 vol. [*Nearly ready.*

FRIENDS in COUNCIL: a Series of Readings and Discourses thereon. 2 vols. fcp. 8vo. 9s.

Friends in Council, SECOND SERIES. 2 vols. post 8vo. 14s.

Essays written in the Intervals of Business. Fcp. 8vo. 2s. 6d.

Companions of My Solitude. By the same Author. Fcp. 8vo. 3s. 6d.

LORD MACAULAY'S MISCELLANEOUS WRITINGS: comprising his Contributions to KNIGHT'S *Quarterly Magazine*, Articles from the Edinburgh Review not included in his *Critical and Historical Essays*, Biographies from the *Encyclopædia Britannica*, Miscellaneous Poems and Inscriptions. 2 vols. 8vo. with Portrait, 21s.

The REV. SYDNEY SMITH'S MISCELLANEOUS WORKS; including his Contributions to the *Edinburgh Review*.

 LIBRARY EDITION. 3 vols. 8vo. 36s.
 TRAVELLER'S EDITION, in 1 vol. 21s.
 In POCKET VOLUMES. 3 vols. 21s.
 PEOPLE'S EDITION. 2 vols. crown 8vo. 8s.

Elementary Sketches of Moral Philosophy, delivered at the Royal Institution. By the same Author. Fcp. 8vo. 7s.

The Wit and Wisdom of Sydney Smith: a Selection of the most memorable Passages in his Writings and Conversation. 16mo. 7s. 6d.

From MATTER to SPIRIT: the Result of Ten Years' Experience in Spirit Manifestations. By C. D. with a preface by A. B. Post 8vo. 8s. 6d.

The HISTORY of the SUPERNATURAL in All Ages and Nations, and in all Churches, Christian and Pagan; Demonstrating a Universal Faith. By WILLIAM HOWITT. 2 vols. post 8vo. 18s.

CHAPTERS on MENTAL PHYSIOLOGY. By Sir HENRY HOLLAND, Bart. M.D. F.R.S. Second Edition. Post 8vo. 8s. 6d.

ESSAYS selected from CONTRIBUTIONS to the *Edinburgh Review*. By HENRY ROGERS. Second Edition. 3 vols. fcp. 21s.

The Eclipse of Faith; or, a Visit to a Religious Sceptic. By the same Author. Tenth Edition. Fcp. 8vo. 5s.

Defence of the Eclipse of Faith, by its Author; a rejoinder to Dr. Newman's *Reply*. Third Edition. Fcp. 8vo. 3s. 6d.

Selections from the Correspondence of R. E. H. Greyson. By the same Author. Third Edition. Crown 8vo. 7s. 6d.

Fulleriana, or the Wisdom and Wit of THOMAS FULLER, with Essay on his Life and Genius. By the same Author. 16mo. 2s. 6d.

Reason and Faith, reprinted from the *Edinburgh Review*. By the same Author. Fourth Edition. Fcp. 8vo. 1s. 6d.

An INTRODUCTION to MENTAL PHILOSOPHY, on the Inductive Method. By J. D. MORELL, M.A. LL.D. 8vo. 12s.

Elements of Psychology, containing the Analysis of the Intellectual Powers. By the same Author. Post 8vo. 7s. 6d.

The SENSES and the INTELLECT. By ALEXANDER BAIN, M.A. Professor of Logic in the University of Aberdeen. Second Edition. 8vo. price 15s.

The Emotions and the Will, by the same Author; completing a Systematic Exposition of the Human Mind. 8vo. 15s.

On the Study of Character, including an Estimate of Phrenology. By the same Author. 8vo. 9s.

HOURS WITH THE MYSTICS: a Contribution to the History of Religious Opinion. By ROBERT ALFRED VAUGHAN, B.A. Second Edition. 2 vols. crown 8vo. 12s.

PSYCHOLOGICAL INQUIRIES, or Essays intended to illustrate the Influence of the Physical Organisation on the Mental Faculties. By Sir B. C. BRODIE, Bart. Fcp. 8vo. 5s. PART II. Essays intended to illustrate some Points in the Physical and Moral History of Man. Fcp. 8vo. 5s.

The PHILOSOPHY of NECESSITY; or Natural Law as applicable to Mental, Moral, and Social Science. By CHARLES BRAY. Second Edition. 8vo. 9s.

The Education of the Feelings and Affections. By the same Author. Third Edition. 8vo. 3s. 6d.

CHRISTIANITY and COMMON SENSE. By Sir WILLOUGHBY JONES, Bart. M.A. Trin. Coll. Cantab. 8vo. 6s.

Astronomy, Meteorology, Popular Geography, &c.

OUTLINES of ASTRONOMY. By Sir J. F. W. HERSCHEL, Bart. M.A. Seventh Edition, revised; with Plates and Woodcuts. 8vo. 18s.

ARAGO'S POPULAR ASTRONOMY. Translated by Admiral W. H. SMYTH, F.R.S. and R. GRANT, M.A. With 25 Plates and 358 Woodcuts. 2 vols. 8vo. £2 5s.

Arago's Meteorological Essays, with Introduction by Baron HUMBOLDT. Translated under the superintendence of Major-General E. SABINE, R.A. 8vo. 18s.

The **WEATHER-BOOK**; a Manual of Practical Meteorology. By Rear-Admiral ROBERT FITZ ROY, R.N. F.R.S. Third Edition, with 16 Diagrams. 8vo. 15s.

SAXBY'S WEATHER SYSTEM, or Lunar Influence on Weather, By S. M. SAXBY, R.N. Principal Instructor of Naval Engineers, H.M. Steam Reserve. Second Edition. Post 8vo. 4s.

DOVE'S LAW of STORMS considered in connexion with the ordinary Movements of the Atmosphere. Translated by R. H. SCOTT, M.A. T.C.D. 8vo. 10s. 6d.

CELESTIAL OBJECTS for COMMON TELESCOPES. By the Rev. T. W. WEBB, M.A. F.R.A.S. With Map of the Moon, and Woodcuts. 16mo. 7s.

PHYSICAL GEOGRAPHY for SCHOOLS and GENERAL READERS. By M. F. MAURY, LL.D. Author of 'Physical Geography of the Sea,' &c. Fcp. 8vo. with 2 Plates, 2s. 6d.

A DICTIONARY, Geographical, Statistical, and Historical, of the various Countries, Places, and Principal Natural Objects in the World. By J. R. M'CULLOCH, Esq. With 6 Maps. 2 vols. 8vo. 63s.

A GENERAL DICTIONARY of GEOGRAPHY, Descriptive, Physical, Statistical, and Historical: forming a complete Gazetteer of the World. By A. KEITH JOHNSTON, F.R.S.E. 8vo. 30s.

A MANUAL of GEOGRAPHY, Physical, Industrial, and Political. By W. HUGHES, F.R.G.S. Professor of Geography in King's College, and in Queen's College, London. With 6 Maps. Fcp. 8vo. 7s. 6d.

Or in Two Parts:—PART I. Europe, 3s. 6d. PART II. Asia, Africa, America, Australasia, and Polynesia, 4s.

The Geography of British History; a Geographical description of the British Islands at Successive Periods, from the Earliest Times to the Present Day. By the same. With 6 Maps. Fcp. 8vo. 8s. 6d.

The **BRITISH EMPIRE**; a Sketch of the Geography, Growth, Natural and Political Features of the United Kingdom, its Colonies and Dependencies. By CAROLINE BRAY. With 5 Maps. Fcp. 8vo. 7s. 6d.

COLONISATION and COLONIES: a Series of Lectures delivered before the University of Oxford. By HERMAN MERIVALE, M.A. Professor of Political Economy. 8vo. 18s.

The **AFRICANS at HOME**: a popular Description of Africa and the Africans. By the Rev. R. M. MACBRAIR, M.A. Second Edition; including an Account of the Discovery of the Source of the Nile. With Map and 70 Woodcuts. Fcp. 8vo. 5s.

MAUNDER'S TREASURY of GEOGRAPHY, Physical, Historical, Descriptive, and Political. Completed by W. HUGHES, F.R.G.S. With 7 Maps and 16 Plates. Fcp. 8vo. 10s.

Natural History and Popular Science.

The **ELEMENTS of PHYSICS or NATURAL PHILOSOPHY.** By NEIL ARNOTT, M.D. F.R.S. Physician Extraordinary to the Queen. Sixth Edition. PART I. 8vo. 10s. 6d.

HEAT CONSIDERED as a MODE of MOTION; a Course of Lectures delivered at the Royal Institution. By Professor JOHN TYNDALL, F.R.S. Crown 8vo. with Woodcuts, 12s. 6d.

VOLCANOS, the Character of their Phenomena, their Share in the Structure and Composition of the Surface of the Globe, &c. By G. POULETT SCROPE, M.P. F.R.S. Second Edition. 8vo. with illustrations, 15s.

A TREATISE on ELECTRICITY, in Theory and Practice. By A. DE LA RIVE, Prof. in the Academy of Geneva. Translated by C. V. WALKER, F.R.S. 3 vols. 8vo. with Woodcuts, £3 13s.

The **CORRELATION of PHYSICAL FORCES.** By W. R. GROVE, Q.C. V.P.R.S. Fourth Edition. 8vo. 7s. 6d.

The **GEOLOGICAL MAGAZINE**; or, Monthly Journal of Geology Edited by T. RUPERT JONES, F.G.S. Professor of Geology in the R. M. College, Sandhurst; assisted by J. C. WOODWARD, F.G.S. F.Z.S. British Museum. 8vo. with Illustrations, price 1s. 6d. monthly.

A GUIDE to GEOLOGY. By J. PHILLIPS, M.A. Professor of Geology in the University of Oxford. Fifth Edition; with Plates and Diagrams. Fcp. 8vo. 4s.

A GLOSSARY of MINERALOGY. By H. W. BRISTOW, F.G.S. of the Geological Survey of Great Britain. With 486 Figures. Crown 8vo. 12s.

PHILLIPS'S ELEMENTARY INTRODUCTION to MINERALOGY, with extensive Alterations and Additions, by H. J. BROOKE, F.R.S. and W. H. MILLER, F.G.S. Post 8vo. with Woodcuts, 18s.

VAN DER HOEVEN'S HANDBOOK of ZOOLOGY. Translated from the Second Dutch Edition by the Rev. W. CLARK, M.D. F.R.S. 2 vols. 8vo. with 24 Plates of Figures, 60s.

The **COMPARATIVE ANATOMY and PHYSIOLOGY of the VERTE**brate Animals. By RICHARD OWEN, F.R.S. D.C.L. 2 vols. 8vo. with upwards of 1,200 Woodcuts. [*In the press.*

HOMES WITHOUT HANDS: an Account of the Habitations constructed by various Animals, classed according to their Principles of Construction. By Rev. J. G. WOOD, M.A. F.L.S. Illustrations on Wood by G. Pearson, from Drawings by F. W. Keyl and E. A. Smith. In course of publication in 20 Parts, 1s. each.

MANUAL of CŒLENTERATA. By J. REAY GREENE, B.A. M.R.I.A. Edited by the Rev. J. A. GALBRAITH, M.A. and the Rev. S. HAUGHTON, M.D. Fcp. 8vo. with 39 Woodcuts. 5s.

Manual of Protozoa; with a General Introduction on the Principles of Zoology. By the same Author and Editors. Fcp. 8vo. with 16 Woodcuts, 2s.

Manual of the Metalloids. By J. APJOHN, M.D. F.R.S. and the same Editors. Fcp. 8vo. with 38 Woodcuts, 7s. 6d.

THE ALPS: Sketches of Life and Nature in the Mountains. By Baron H. VON BERLEPSCH. Translated by the Rev. L. STEPHEN, M.A. With 17 Illustrations. 8vo. 15s.

The **SEA** and its **LIVING WONDERS**. By Dr. G. HARTWIG. Second (English) Edition. 8vo. with many Illustrations. 18s.

The **TROPICAL WORLD**. By the same Author. With 8 Chromoxylographs and 172 Woodcuts. 8vo. 21s.

SKETCHES of the **NATURAL HISTORY** of **CEYLON**. By Sir J. EMERSON TENNENT, K.C.S. LL.D. With 82 Wood Engravings. Post 8vo. price 12s. 6d.

Ceylon. By the same Author. 5th Edition; with Maps, &c. and 90 Wood Engravings. 2 vols. 8vo. £2 10s.

MARVELS and **MYSTERIES** of **INSTINCT**; or, Curiosities of Animal Life. By G. GARRATT. Third Edition. Fcp. 8vo. 7s.

HOME WALKS and **HOLIDAY RAMBLES**. By the Rev. C. A. JOHNS, B.A. F.L.S. Fcp. 8vo. with 10 Illustrations, 6s.

KIRBY and **SPENCE'S INTRODUCTION** to **ENTOMOLOGY**, or Elements of the Natural History of Insects. Seventh Edition. Crown 8vo. price 5s.

MAUNDER'S TREASURY of **NATURAL HISTORY**, or Popular Dictionary of Zoology. Revised and corrected by T. S. COBBOLD, M.D. Fcp. 8vo. with 900 Woodcuts, 10s.

The **TREASURY** of **BOTANY**, on the Plan of Maunder's Treasury. By J. LINDLEY, M.D. and T. MOORE, F.L.S. assisted by other Practical Botanists. With 16 Plates, and many Woodcuts from designs by W. H. Fitch. Fcp. 8vo. *[In the press.*

The **ROSE AMATEUR'S GUIDE**. By THOMAS RIVERS. 8th Edition. Fcp. 8vo. 4s.

The **BRITISH FLORA**; comprising the Phænogamous or Flowering Plants and the Ferns. By Sir W. J. HOOKER, K.H. and G. A. WALKER ARNOTT, LL.D. 12mo. with 12 Plates, 14s. or coloured, 21s.

BRYOLOGIA BRITANNICA; containing the Mosses of Great Britain and Ireland, arranged and described. By W. WILSON. 8vo. with 61 Plates 42s. or coloured, £4 4s.

The **INDOOR GARDENER**. By Miss MALING. Fcp. 8vo. with coloured Frontispiece, 5s.

LOUDON'S ENCYCLOPÆDIA of **PLANTS**; comprising the Specific Character, Description, Culture, History, &c. of all the Plants found in Great Britain. With upwards of 12,000 Woodcuts. 8vo. £3 13s. 6d.

Loudon's Encyclopædia of Trees and Shrubs; containing the Hardy Trees and Shrubs of Great Britain scientifically and popularly described. With 2,000 Woodcuts 8vo. 50s.

HISTORY of the **BRITISH FRESHWATER ALGÆ**. By A. H. HASSALL, M.D. With 100 Plates of Figures. 2 vols. 8vo. price £1 15s.

MAUNDER'S SCIENTIFIC and LITERARY TREASURY; a Popular Encyclopædia of Science, Literature, and Art. Fcp. 8vo. 10s.

A DICTIONARY of SCIENCE, LITERATURE and ART; comprising the History, Description, and Scientific Principles of every Branch of Human Knowledge. Edited by W. T. BRANDE, F.R.S.L. and E. Fourth Edition, revised and corrected. [*In the press.*

ESSAYS on SCIENTIFIC and other SUBJECTS, contributed to the *Edinburgh* and *Quarterly Reviews*. By Sir H. HOLLAND, Bart. M.D. Second Edition. 8vo. 14s.

ESSAYS from the EDINBURGH and QUARTERLY REVIEWS; with Addresses and other pieces. By Sir J. F. W. HERSCHEL, Bart. M.A. 8vo. 18s.

Chemistry, Medicine, Surgery, and the Allied Sciences.

A DICTIONARY of CHEMISTRY and the Allied Branches of other Sciences; founded on that of the late Dr. Ure. By HENRY WATTS, F.C.S. assisted by eminent Contributors. 4 vols. 8vo. in course of publication in Monthly Parts. VOL. I. 31s. 6d. and VOL. II. 26s. are now ready.

HANDBOOK of CHEMICAL ANALYSIS, adapted to the Unitary System of Notation: Based on Dr. H. Wills' *Anleitung zur chemischen Analyse*. By F. T. CONINGTON, M.A. F.C.S. Post 8vo. 7s. 6d.—TABLES of QUALITATIVE ANALYSIS to accompany the same, 2s. 6d.

A HANDBOOK of VOLUMETRICAL ANALYSIS. By ROBERT H. SCOTT, M.A. T.C.D. Post 8vo. 4s. 6d.

ELEMENTS of CHEMISTRY, Theoretical and Practical. By WILLIAM A. MILLER, M.D. LL.D. F.R.S. F.G.S. Professor of Chemistry, King's College, London. 3 vols. 8vo. £2 12s. PART I. CHEMICAL PHYSICS. Third Edition enlarged, 12s. PART II. INORGANIC CHEMISTRY. Second Edition, 20s. PART III. ORGANIC CHEMISTRY. Second Edition, 20s.

A MANUAL of CHEMISTRY, Descriptive and Theoretical. By WILLIAM ODLING, M.B. F.R.S. Lecturer on Chemistry at St. Bartholomew's Hospital. PART I. 8vo. 9s.

A Course of Practical Chemistry, for the use of Medical Students. By the same Author. PART I. crown 8vo. with Woodcuts, 4s. 6d. PART II. (completion) *just ready*.

The DIAGNOSIS and TREATMENT of the DISEASES of WOMEN; including the Diagnosis of Pregnancy. By GRAILY HEWITT, M.D. Physician to the British Lying-in Hospital. 8vo. 16s.

LECTURES on the DISEASES of INFANCY and CHILDHOOD. By CHARLES WEST, M.D. &c. Fourth Edition, revised and enlarged. 8vo. 14s.

EXPOSITION of the SIGNS and SYMPTOMS of PREGNANCY: with other Papers on subjects connected with Midwifery. By W. F. MONTGOMERY, M.A. M.D. M.R.I.A. 8vo. with Illustrations, 25s.

A SYSTEM of SURGERY, Theoretical and Practical. In Treatises by Various Authors, arranged and edited by T. HOLMES, M.A. Cantab. Assistant-Surgeon to St. George's Hospital. 4 vols. 8vo.

Vol. I. **General Pathology.** 21s.

Vol. II. **Local Injuries—Diseases of the Eye.** 21s.

Vol. III. **Operative Surgery. Diseases of the Organs of Special** Sense, Respiration, Circulation, Locomotion and Innervation. 21s.

Vol. IV. **Diseases of the Alimentary Canal, of the Urino-genitary** Organs, of the Thyroid, Mamma and Skin; with Appendix of Miscellaneous Subjects, and GENERAL INDEX. *[Early in October.*

LECTURES on the PRINCIPLES and PRACTICE of PHYSIC. By THOMAS WATSON, M.D. Physician-Extraordinary to the Queen. Fourth Edition. 2 vols. 8vo. 34s.

LECTURES on SURGICAL PATHOLOGY. By J. PAGET, F.R.S. Surgeon-Extraordinary to the Queen. Edited by W. TURNER, M.B. 8vo. with 117 Woodcuts, 21s.

A TREATISE on the CONTINUED FEVERS of GREAT BRITAIN. By C. MURCHISON, M.D. Senior Physician to the London Fever Hospital. 8vo. with coloured Plates, 18s.

DEMONSTRATIONS of MICROSCOPIC ANATOMY; a Guide to the Examination of the Animal Tissues and Fluids in Health and Disease, for the use of the Medical and Veterinary Professions. Founded on a Course of Lectures delivered by Dr. HARLEY, Prof. in Univ. Coll. London. Edited by G. T. BROWN, late Vet. Prof. in the Royal Agric. Coll. Cirencester. 8vo. with Illustrations. *[Nearly ready.*

ANATOMY, DESCRIPTIVE and SURGICAL. By HENRY GRAY, F.R.S. With 410 Wood Engravings from Dissections. Third Edition, by T. HOLMES, M.A. Cantab. Royal 8vo. 28s.

PHYSIOLOGICAL ANATOMY and PHYSIOLOGY of MAN. By the late R. B. TODD, M.D. F.R.S. and W. BOWMAN, F.R.S. of King's College. With numerous Illustrations. VOL. II. 8vo. 25s.

A New Edition of the FIRST VOLUME, by Dr. LIONEL S. BEALE, is preparing for publication.

The CYCLOPÆDIA of ANATOMY and PHYSIOLOGY. Edited by the late R. B. TODD, M.D. F.R.S. Assisted by nearly all the most eminent cultivators of Physiological Science of the present age. 5 vols. 8vo. with 2,853 Woodcuts, £6 6s.

A DICTIONARY of PRACTICAL MEDICINE. By J. COPLAND, M.D. F.R.S. Abridged from the larger work by the Author, assisted by J. C. COPLAND. 1 vol. 8vo. *[In the press.*

Dr. Copland's Dictionary of Practical Medicine (the larger work). 3 vols. 8vo. £5 11s.

The WORKS of SIR B. C. BRODIE, Bart. Edited by CHARLES HAWKINS, F.R.C.S.E. 2 vols. 8vo. *[In the press.*

MEDICAL NOTES and REFLECTIONS. By Sir H. HOLLAND, Bart. M.D. Third Edition. 8vo. 18s.

HOOPER'S MEDICAL DICTIONARY, or Encyclopædia of Medical Science. Ninth Edition, brought down to the present time, by ALEX. HENRY, M.D. 1 vol. 8vo. [*In the press.*

A MANUAL of MATERIA MEDICA and THERAPEUTICS, abridged from Dr. PEREIRA'S *Elements* by F. J. FARRE, M.D. Cantab. assisted by R. BENTLEY, M.R.C.S. and by R. WARRINGTON, F.C.S. 1 vol. 8vo.

Dr. Pereira's Elements of Materia Medica and Therapeutics, Third Edition. By A. S. TAYLOR, M.D. and G. O. REES, M.D. 3 vols. 8vo. with numerous Woodcuts, £3 15s.

The Fine Arts, and Illustrated Editions.

The **NEW TESTAMENT of OUR LORD** and **SAVIOUR JESUS CHRIST.** Illustrated with numerous Engravings on Wood from the OLD MASTERS. Crown 4to. price 63s. cloth, gilt top; or price £5 5s. elegantly bound in morocco. [*In October.*

LYRA GERMANICA; Hymns for the Sundays and Chief Festivals of the Christian Year. Translated by CATHERINE WINKWORTH: 125 Illustrations on Wood drawn by J. LEIGHTON, F.S.A. Fcp. 4to. 21s.

CATS' and FARLIE'S MORAL EMBLEMS; with Aphorisms, Adages, and Proverbs of all Nations: comprising 121 Illustrations on Wood by J. LEIGHTON, F.S.A. with an appropriate Text by R. PIGOTT. Imperial 8vo. 31s. 6d.

BUNYAN'S PILGRIM'S PROGRESS: with 126 Illustrations on Steel and Wood by C. BENNETT; and a Preface by the Rev. C. KINGSLEY. Fcp. 4to. 21s.

The **HISTORY of OUR LORD**, as exemplified in Works of Art: with that of His Types, St. John the Baptist, and other Persons of the Old and New Testament. By Mrs. JAMESON and Lady EASTLAKE. Being the Fourth and concluding SERIES of 'Sacred and Legendary Art;' with 81 Etchings and 281 Woodcuts. 2 vols. square crown 8vo. 42s.

In the same Series, by Mrs. JAMESON.

Legends of the Saints and Martyrs. Fourth Edition, with 19 Etchings and 187 Woodcuts. 2 vols. 31s. 6d.

Legends of the Monastic Orders. Third Edition, with 11 Etchings and 88 Woodcuts. 1 vol. 21s.

Legends of the Madonna. Third Edition, with 27 Etchings and 165 Woodcuts. 1 vol. 21s.

Arts, Manufactures, &c.

ENCYCLOPÆDIA of ARCHITECTURE, Historical, Theoretical, and Practical. By JOSEPH GWILT. With more than 1,000 Woodcuts. 8vo. 42s.

TUSCAN SCULPTORS, their Lives, Works, and Times. With Illustrations from Original Drawings and Photographs. By CHARLES C. PERKINS. 2 vols. imperial 8vo. [*In the press.*

The **ENGINEER'S HANDBOOK**; explaining the Principles which should guide the young Engineer in the Construction of Machinery. By C. S. LOWNDES. Post 8vo. 5s.

The **ELEMENTS of MECHANISM**, for Students of Applied Mechanics. By T. M. GOODEVE, M.A. Professor of Nat. Philos. in King's Coll. London. With 206 Woodcuts. Post 8vo. 6s. 6d.

URE'S DICTIONARY of ARTS, MANUFACTURES, and MINES. Re-written and enlarged by ROBERT HUNT, F.R.S. assisted by numerous gentlemen eminent in Science and the Arts. With 2,000 Woodcuts. 3 vols. 8vo. £4.

ENCYCLOPÆDIA of CIVIL ENGINEERING, Historical, Theoretical, and Practical. By E. CRESY, C.E. With above 3,000 Woodcuts. 8vo. 42s.

TREATISE on MILLS and MILLWORK. By W. FAIRBAIRN, C.E. F.R.S. With 18 Plates and 322 Woodcuts. 2 vols. 8vo. 32s. or each vol. separately, 16s.

Useful Information for Engineers. By the same Author. FIRST and SECOND SERIES, with many Plates and Woodcuts. 2 vols. crown 8vo. 21s. or each vol. separately, 10s. 6d.

The **Application of Cast and Wrought Iron to Building Purposes.** By the same Author. Third Edition, with Plates and Woodcuts. [*Nearly ready.*

The **PRACTICAL MECHANIC'S JOURNAL**: An Illustrated Record of Mechanical and Engineering Science, and Epitome of Patent Inventions. 4to. price 1s. monthly.

The **PRACTICAL DRAUGHTSMAN'S BOOK of INDUSTRIAL DESIGN.** By W. JOHNSON, Assoc. Inst. C.E. With many hundred Illustrations. 4to. 28s. 6d.

The **PATENTEE'S MANUAL**; a Treatise on the Law and Practice of Letters Patent for the use of Patentees and Inventors. By J. and J. H. JOHNSON. Post 8vo. 7s. 6d.

The **ARTISAN CLUB'S TREATISE on the STEAM ENGINE**, in its various Applications to Mines, Mills, Steam Navigation, Railways and Agriculture. By J. BOURNE, C.E. Fifth Edition; with 37 Plates and 546 Woodcuts. 4to. 42s.

A Catechism of the Steam Engine, in its various Applications to Mines, Mills, Steam Navigation, Railways, and Agriculture. By the same Author. With 80 Woodcuts. Fcp. 8vo. 6s.

The **STORY of the GUNS.** By Sir J. EMERSON TENNENT, K.C.S. F.R.S. With 33 Woodcuts. Post 8vo. 7s. 6d.

The **THEORY of WAR** Illustrated by numerous Examples from History. By Lieut.-Col. P. L. MACDOUGALL. *Third Edition*, with 10 Plans. Post 8vo. 10s. 6d.

COLLIERIES and COLLIERS; A Handbook of the Law and leading. Cases relating thereto. By J. C. FOWLER, Barrister-at-Law. Fcp. 8vo. 6s.

The **ART of PERFUMERY**; the History and Theory of Odours, and the Methods of Extracting the Aromas of Plants. By Dr. PIESSE, F.C.S. Third Edition, with 53 Woodcuts. Crown 8vo. 10s. 6d.

Chemical, Natural, and Physical Magic, for Juveniles during the Holidays. By the same Author. With 30 Woodcuts. Fcp. 8vo. 3s. 6d.

The **Laboratory of Chemical Wonders**: a Scientific Mélange for Young People. By the same. Crown 8vo. 5s. 6d.

TALPA; or the Chronicles of a Clay Farm. By C. W. HOSKYNS, Esq. With 24 Woodcuts from Designs by G. CRUIKSHANK. 16mo. 5s. 6d.

H.R.H. the PRINCE CONSORT'S FARMS: An Agricultural Memoir. By JOHN CHALMERS MORTON. Dedicated by permission to Her Majesty the QUEEN. With 40 Wood Engravings. 4to. 52s. 6d.

Handbook of Farm Labour, Steam, Water, Wind, Horse Power, Hand Power, &c. By the same Author. 16mo. 1s. 6d.

Handbook of Dairy Husbandry; comprising the General Management of a Dairy Farm, &c. By the same. 16mo. 1s. 6d.

LOUDON'S ENCYCLOPÆDIA of AGRICULTURE: comprising the Laying-out, Improvement, and Management of Landed Property, and the Cultivation and Economy of the Productions of Agriculture. With 1,100 Woodcuts. 8vo. 31s. 6d.

Loudon's Encylopædia of Gardening: Comprising the Theory and Practice of Horticulture, Floriculture, Arboriculture, and Landscape Gardening. With 1,000 Woodcuts. 8vo. 31s. 6d.

Loudon's Encyclopædia of Cottage, Farm, and Villa Architecture and Furniture. With more than 2,000 Woodcuts. 8vo. 42s.

HISTORY of WINDSOR GREAT PARK and WINDSOR FOREST. By WILLIAM MENZIES, Resident Deputy Surveyor. Dedicated by permission to H. M. the QUEEN. With 2 Maps, and 20 Photographs by the EARL of CAITHNESS and Mr. BEMBRIDGE. Imperial folio, £8 8s.

BAYLDON'S ART of VALUING RENTS and TILLAGES, and Claims of Tenants upon Quitting Farms, both at Michaelmas and Lady-Day. Eighth Edition, adapted to the present time by J. C. MORTON.

Religious and Moral Works.

An **EXPOSITION of the 39 ARTICLES**, Historical and Doctrinal. By E. HAROLD BROWNE, D.D. Lord Bishop of Ely. Sixth Edition, 8vo. 16s.

The Pentateuch and the Elohistic Psalms, in reply to Bishop Colenso. By the same Author. 8vo. 2s.

Examination Questions on Bishop Browne's Exposition of the Articles. By the Rev. J. GORLE, M.A. Fcp. 3s. 6d.

FIVE LECTURES on the CHARACTER of ST. PAUL; being the Hulsean Lectures for 1862. By the Rev. J. S. HOWSON, D.D. Second Edition. 8vo. 9s.

A CRITICAL and GRAMMATICAL COMMENTARY on ST. PAUL'S Epistles. By C. J. ELLICOTT, D.D. Lord Bishop of Gloucester and Bristol. 8vo.

Galatians, Third Edition, 8s. 6d.

Ephesians, Third Edition, 8s. 6d.

Pastoral Epistles, Second Edition, 10s. 6d.

Philippians, Colossians, and Philemon, Second Edition, 10s. 6d.

Thessalonians, Second Edition, 7s. 6d.

Historical Lectures on the Life of our Lord Jesus Christ: being the Hulsean Lectures for 1859. By the same. Third Edition. 8vo. 10s. 6d.

The Destiny of the Creature; and other Sermons preached before the University of Cambridge. By the same. Post 8vo. 5s.

The Broad and the Narrow Way; Two Sermons preached before the University of Cambridge. By the same. Crown 8vo. 2s.

Rev. T. H. HORNE'S INTRODUCTION to the CRITICAL STUDY and Knowledge of the Holy Scriptures. Eleventh Edition, corrected and extended under careful Editorial revision. With 4 Maps and 22 Woodcuts and Facsimiles. 4 vols. 8vo. £3 13s. 6d.

Rev. T. H. Horne's Compendious Introduction to the Study of the Bible, being an Analysis of the larger work by the same Author. Re-edited by the Rev. JOHN AYRE, M.A. With Maps. &c. Post 8vo. 9s.

The TREASURY of BIBLE KNOWLEDGE, on the Plan of Maunder's Treasuries. By the Rev. JOHN AYRE, M.A. Fcp. 8vo. with Maps and Illustrations. *[In the press.*

The GREEK TESTAMENT; with Notes, Grammatical and Exegetical. By the Rev. W. WEBSTER, M.A. and the Rev. W. F. WILKINSON, M.A. 2 vols. 8vo. £2 4s.

VOL. I. the Gospels and Acts, 20s.

VOL. II. the Epistles and Apocalypse, 24s.

The FOUR EXPERIMENTS in Church and State; and the Conflicts of Churches. By Lord ROBERT MONTAGU, M.P. 8vo. 12s.

EVERY-DAY SCRIPTURE DIFFICULTIES explained and illustrated; Gospels of St. Matthew and St. Mark. By J. E. PRESCOTT, M.A. late Fellow of C. C. Coll. Cantab. 8vo. 9s.

The PENTATEUCH and BOOK of JOSHUA Critically Examined. By J. W. COLENSO, D.D. Lord Bishop of Natal. PART I. *the Pentateuch examined as an Historical Narrative.* 8vo. 6s. PART II. *the Age and Authorship of the Pentateuch Considered,* 7s. 6d. PART III. *the Book of Deuteronomy,* 8s. PART IV. *the First 11 Chapters of Genesis examined and separated, with Remarks on the Creation, the Fall, and the Deluge,* 10s. 6d.

The LIFE and EPISTLES of ST. PAUL. By W. J. CONYBEARE, M.A. late Fellow of Trin. Coll. Cantab. and J. S. HOWSON, D.D. Principal of the Collegiate Institution, Liverpool.

LIBRARY EDITION, with all the Original Illustrations, Maps, Landscapes on Steel, Woodcuts, &c. 2 vols. 4to. 48s.

INTERMEDIATE EDITION, with a Selection of Maps, Plates, and Woodcuts. 2 vols. square crown 8vo. 31s. 6d.

PEOPLE'S EDITION, revised and condensed, with 46 Illustrations and Maps. 2 vols. crown 8vo. 12s.

The VOYAGE and SHIPWRECK of ST. PAUL; with Dissertations on the Ships and Navigation of the Ancients. By JAMES SMITH, F.R.S. Crown 8vo. Charts, 8s. 6d.

HIPPOLYTUS and his AGE; or, the Beginnings and Prospects of Christianity. By Baron BUNSEN, D.D. 2 vols. 8vo. 30s.

Outlines of the Philosophy of Universal History, applied to Language and Religion: Containing an Account of the Alphabetical Conferences. By the same Author. 2 vols. 8vo. 33s.

Analecta Ante-Nicæna. By the same Author. 3 vols. 8vo. 42s.

THEOLOGIA GERMANICA. Translated by SUSANNAH WINKWORTH: with a Preface by the Rev. C. KINGSLEY; and a Letter by Baron BUNSEN. Fcp. 8vo. 5s.

INSTRUCTIONS in the DOCTRINE and PRACTICE of CHRIS-tianity, as an Introduction to Confirmation. By G. E. L. COTTON, D.D. Lord Bishop of Calcutta. 18mo. 2s. 6d.

ESSAYS on RELIGION and LITERATURE. By Cardinal WISEMAN, Dr. D. ROCK, F. H. LAING, and other Writers. Edited by H. E. MANNING, D.D. 8vo.

ESSAYS and REVIEWS. By the Rev. W. TEMPLE, D.D. the Rev. R. WILLIAMS, B.D. the Rev. B. POWELL, M.A. the Rev. H. B. WILSON, B.D. C. W. GOODWIN, M.A. the Rev. M. PATTISON, B.D. and the Rev. B. JOWETT, M.A. 11th Edition. Fcp. 8vo. 5s.

MOSHEIM'S ECCLESIASTICAL HISTORY. MURDOCK and SOAMES's Translation and Notes, re-edited by the Rev. W. STUBBS, M.A. 3 vols. 8vo. 45s.

The GENTILE and the JEW in the Courts of the Temple of Christ: an Introduction to the History of Christianity. From the German of Prof. DÖLLINGER, by the Rev. N. DARNELL, M.A. 2 vols. 8vo. 21s.

PHYSICO-PROPHETICAL ESSAYS, on the Locality of the Eternal Inheritance, its Nature and Character; the Resurrection Body; and the Mutual Recognition of Glorified Saints. By the Rev. W. LISTER, F.G.S. Crown 8vo. 6s.

BISHOP JEREMY TAYLOR'S ENTIRE WORKS: With Life by BISHOP HEBER. Revised and corrected by the Rev. C. P. EDEN, 10 vols. 8vo. £5 5s.

PASSING THOUGHTS on RELIGION. By the Author of 'Amy Herbert.' 8th Edition. Fcp. 8vo. 5s.

Thoughts for the Holy Week, for Young Persons. By the same Author. 2d Edition. Fcp. 8vo. 2s.

Night Lessons from Scripture. By the same Author. 2d Edition. 32mo. 3s.

Self-Examination before Confirmation. By the same Author. 32mo. price 1s. 6d.

Readings for a Month Preparatory to Confirmation, from Writers of the Early and English Church. By the same. Fcp. 4s.

Readings for Every Day in Lent, compiled from the Writings of Bishop JEREMY TAYLOR. By the same. Fcp. 8vo. 5s.

Preparation for the Holy Communion; the Devotions chiefly from the works of JEREMY TAYLOR. By the same. 32mo. 3s.

MORNING CLOUDS. Second Edition. Fcp. 8vo. 5s.

The Afternoon of Life. By the same Author. Second Edition. Fcp. 5s.

Problems in Human Nature. By the same. Post 8vo. 5s.

The WIFE'S MANUAL; or, Prayers, Thoughts, and Songs on Several Occasions of a Matron's Life. By the Rev. W. CALVERT, M.A. Crown 8vo. price 10s. 6d.

SPIRITUAL SONGS for the SUNDAYS and HOLIDAYS throughout the Year. By J. S. B. MONSELL, LL.D. Vicar of Egham. Third Edition. Fcp. 8vo. 4s. 6d.

HYMNOLOGIA CHRISTIANA; or, Psalms and Hymns selected and arranged in the order of the Christian Seasons. By B. H. KENNEDY, D.D. Prebendary of Lichfield. Crown 8vo. 7s. 6d.

LYRA SACRA; Hymns, Ancient and Modern, Odes and Fragments of Sacred Poetry. Edited by the Rev. B. W. SAVILE, M.A. Fcp. 8vo. 5s.

LYRA GERMANICA, translated from the German by Miss C. WINKWORTH. FIRST SERIES, Hymns for the Sundays and Chief Festivals; SECOND SERIES, the Christian Life. Fcp. 8vo. 5s. each SERIES.

Hymns from Lyra Germanica, 18mo. 1s.

LYRA EUCHARISTICA; Hymns and Verses on the Holy Communion, Ancient and Modern: with other Poems. Edited by the Rev. ORBY SHIPLEY, M.A. Second Edition, revised and enlarged. Fcp. 8vo. 7s. 6d.

Lyra Messianica; Hymns and Verses on the Life of Christ, Ancient and Modern; with other Poems. By the same Editor. Fcp. 8vo. 7s. 6d.

Lyra Mystica; Hymns and Verses on Sacred Subjects, Ancient and Modern. Forming a companion volume to the above, by the same Editor. Fcp. 8vo. [*Nearly ready.*

LYRA DOMESTICA; Christian Songs for Domestic Edification. Translated from the *Psaltery and Harp* of C. J. P. SPITTA, and from other sources, by RICHARD MASSIE. FIRST and SECOND SERIES, fcp. 8vo. price 4s. 6d. each.

The CHORALE BOOK for ENGLAND; a complete Hymn-Book in accordance with the Services and Festivals of the Church of England: the Hymns translated by Miss C. WINKWORTH; the tunes arranged by Prof. W. S. BENNETT and OTTO GOLDSCHMIDT. Fcp. 4to. 10s. 6d.

Congregational Edition. Fcp. 8vo. price 1s. 6d.

Travels, Voyages, &c.

EASTERN EUROPE and WESTERN ASIA. Political and Social Sketches on Russia, Greece, and Syria. By HENRY A. TILLEY. With 6 Illustrations. Post 8vo. 10s. 6d.

EXPLORATIONS in SOUTH-WEST AFRICA, from Walvisch Bay to Lake Ngami and the Victoria Falls. By THOMAS BAINES. 8vo. with Map and Illustrations. [*In October.*

SOUTH AMERICAN SKETCHES; or, a Visit to Rio Janeiro, the Organ Mountains, La Plata, and the Paraná. By THOMAS W. HINCHLIFF, M.A. F.R.G.S. Post 8vo. with Illustrations, 12s. 6d.

EXPLORATIONS in LABRADOR. By HENRY Y. HIND, M.A. F.R.G.S. With Maps and Illustrations. 2 vols. 8vo. 32s.

The Canadian Red River and Assinniboine and Saskatchewan Exploring Expeditions. By the same Author. With Maps and Illustrations. 2 vols. 8vo. 42s.

The CAPITAL of the TYCOON; a Narrative of a Three Years' Residence in Japan. By Sir RUTHERFORD ALCOCK, K.C.B. 2 vols. 8vo. with numerous Illustrations, 42s.

LAST WINTER in ROME and other ITALIAN CITIES. By C. R. WELD, Author of 'The Pyrenees, West and East,' &c. 1 vol. post 8vo. with a Portrait of 'STELLA,' and Engravings on Wood from Sketches by the Author. [*In the Autumn.*

AUTUMN RAMBLES in NORTH AFRICA. By JOHN ORMSBY, of the Middle Temple, Author of the 'Ascent of the Grivola,' in 'Peaks, Passes, and Glaciers.' With 13 Illustrations on Wood from Sketches by the Author. Post 8vo. 8s. 6d.

PEAKS, PASSES, and GLACIERS; a Series of Excursions by Members of the Alpine Club. Edited by J. BALL, M.R.I.A. Fourth Edition; Maps, Illustrations, Woodcuts. Square crown 8vo. 21s.—TRAVELLERS' EDITION, condensed, 16mo. 5s. 6d.

Second Series, edited by E. S. KENNEDY, M.A. F.R.G.S. With many Maps and Illustrations. 2 vols. square crown 8vo. 42s.

Nineteen Maps of the Alpine Districts, from the First and Second Series of *Peaks, Passes, and Glaciers*. Price 7s. 6d.

The **DOLOMITE MOUNTAINS**. Excursions through Tyrol, Carinthia, Carniola, and Friuli in 1861, 1862, and 1863. By J. GILBERT and G. C. CHURCHILL, F.R.G.S. With numerous Illustrations. Square crown 8vo. 21s.

MOUNTAINEERING in 1861; a Vacation Tour. By Prof. J. TYNDALL, F.R.S. Square crown 8vo. with 2 Views, 7s. 6d.

A SUMMER TOUR in the GRISONS and ITALIAN VALLEYS of the Bernina. By Mrs. HENRY FRESHFIELD. With 2 Coloured Maps and 4 Views. Post 8vo. 10s. 6d.

Alpine Byeways; or, Light Leaves gathered in 1859 and 1860. By the same Authoress. Post 8vo. with Illustrations, 10s. 6d.

A LADY'S TOUR ROUND MONTE ROSA; including Visits to the Italian Valleys. With Map and Illustrations. Post 8vo. 14s.

GUIDE to the **PYRENEES**, for the use of Mountaineers. By CHARLES PACKE. With Maps, &c. and a new Appendix. Fcp. 6s.

GUIDE to the **CENTRAL ALPS**, including the Bernese Oberland, Eastern Switzerland, Lombardy, and Western Tyrol. By JOHN BALL, M.R.I.A. Post 8vo. with 8 Maps, 7s. 6d. or with an INTRODUCTION on Alpine Travelling, and on the Geology of the Alps, 8s. 6d. The INTRODUCION separately, 1s.

Guide to the Western Alps. By the same Author. With an Article on the Geology of the Alps by M. E. DESOR. Post 8vo. with Maps, &c. 7s. 6d.

A WEEK at the LAND'S END. By J. T. BLIGHT; assisted by E. H. RODD, R. Q. COUCH, and J. RALFS. With Map and 96 Woodcuts. Fcp. 8vo. 6s. 6d.

VISITS to REMARKABLE PLACES: Old Halls, Battle-Fields, and Scenes Illustrative of Striking Passages in English History and Poetry. By WILLIAM HOWITT. 2 vols. square crown 8vo. with Wood Engravings, price 25s.

The **RURAL LIFE of ENGLAND**. By the same Author. With Woodcuts by Bewick and Williams. Medium 8vo. 12s. 6d.

Works of Fiction.

LATE LAURELS: a Tale. By the Author of 'Wheat and Tares.' 2 vols. post 8vo. 15s.

GRYLL GRANGE. By the Author of 'Headlong Hall.' Post 8vo. price 7s. 6d.

A FIRST FRIENDSHIP. [Reprinted from *Fraser's Magazine.*] Crown 8vo. 7s. 6d.

THALATTA; or, the Great Commoner : a Political Romance. Crown 8vo. 9s.

ATHERSTONE PRIORY. By L. N. COMYN. 2 vols. post 8vo. 21s.

Ellice: a Tale. By the same. Post 8vo. 9s. 6d.

The LAST of the OLD SQUIRES. By the Rev. J. W. WARTER, B.D. Second Edition. Fcp. 8vo. 4s. 6d.

TALES and STORIES by the Author of 'Amy Herbert,' uniform Edition, each Story *or* Tale in a single Volume.

AMY HERBERT, 2s. 6d.	IVORS, 3s. 6d.
GERTRUDE, 2s. 6d.	KATHARINE ASHTON, 3s. 6d.
EARL'S DAUGHTER, 2s. 6d.	MARGARET PERCIVAL, 5s.
EXPERIENCE OF LIFE, 2s. 6d.	LANETON PARSONAGE, 4s. 6d.
CLEVE HALL, 3s. 6d.	URSULA, 4s. 6d.

A Glimpse of the World. By the Author of 'Amy Herbert.' Fcp. 7s. 6d.

ESSAYS on FICTION; comprising Articles on Sir W. SCOTT, Sir E. B. LYTTON, Colonel SENIOR, Mr. THACKERAY, and Mrs. BEECHER STOWE. Reprinted chiefly from the *Edinburgh, Quarterly*, and *Westminster Reviews*; with large Additions. By NASSAU W. SENIOR. Post 8vo. 10s. 6d.

The GLADIATORS: A Tale of Rome and Judæa. By G. J. WHYTE MELVILLE. Crown 8vo.

Digby Grand, an Autobiography. By the same Author. 1 vol. 5s.

Kate Coventry, an Autobiography. By the same. 1 vol. 5s.

General Bounce, or the Lady and the Locusts. By the same. 1 vol. 5s.

Holmby House, a Tale of Old Northamptonshire. 1 vol. 5s.

Good for Nothing, or All Down Hill. By the same. 1 vol. 6s.

The Queen's Maries, a Romance of Holyrood. 1 vol. 6s.

The Interpreter, a Tale of the War. By the same. 1 vol. 5s.

TALES from GREEK MYTHOLOGY. By the Rev. G. W. COX, M.A. late Scholar of Trin. Coll. Oxon. Second Edition. Square 16mo. 3s. 6d.

Tales of the Gods and Heroes. By the same Author. Second Edition. Fcp. 8vo. 5s.

Tales of Thebes and Argos. By the same Author. Fcp. 8vo. 4s. 6d.

The WARDEN: a Novel. By ANTHONY TROLLOPE. Crown 8vo. 3s. 6d.

Barchester Towers: a Sequel to 'The Warden.' By the same Author. Crown 8vo. 5s.

The SIX SISTERS of the VALLEYS: an Historical Romance. By W. BRAMLEY-MOORE, M.A. Incumbent of Gerrard's Cross, Bucks. With 14 Illustrations on Wood. Crown 8vo. 5s.

Poetry and the Drama.

MOORE'S POETICAL WORKS, Cheapest Editions complete in 1 vol. including the Autobiographical Prefaces and Author's last Notes, which are still copyright. Crown 8vo. ruby type, with Portrait, 7s. 6d. or People's Edition, in larger type, 12s. 6d.

Moore's Poetical Works, as above, Library Edition, medium 8vo. with Portrait and Vignette, 21s. or in 10 vols. fcp. 3s. 6d. each.

TENNIEL'S EDITION of MOORE'S LALLA ROOKH, with 68 Wood Engravings from original Drawings and other Illustrations. Fcp. 4to. 21s.

Moore's Lalla Rookh. 32mo. Plate, 1s. 16mo. Vignette, 2s. 6d. Square crown 8vo. with 13 Plates, 15s.

MACLISE'S EDITION of MOORE'S IRISH MELODIES, with 161 Steel Plates from Original Drawings. Super-royal 8vo. 31s. 6d.

Moore's Irish Melodies, 32mo. Portrait, 1s. 16mo. Vignette, 2s. 6d. Square crown 8vo. with 13 Plates, 21s.

SOUTHEY'S POETICAL WORKS, with the Author's last Corrections and copyright Additions. Library Edition, in 1 vol. medium 8vo. with Portrait and Vignette, 14s. or in 10 vols. fcp. 3s. 6d. each.

LAYS of ANCIENT ROME; with *Ivry* and the *Armada*. By the Right Hon. LORD MACAULAY. 16mo. 4s. 6d.

Lord Macaulay's Lays of Ancient Rome. With 90 Illustrations on Wood, Original and from the Antique, from Drawings by G. SCHARF. Fcp. 4to. 21s.

POEMS. By JEAN INGELOW. Seventh Edition. Fcp. 8vo. 5s.

POETICAL WORKS of LETITIA ELIZABETH LANDON (L. E. L.) 2 vols. 16mo 10s.

PLAYTIME with the POETS: a Selection of the best English Poetry for the use of Children. By a LADY. Crown 8vo. 5s.

The REVOLUTIONARY EPICK. By the Right Hon. BENJAMIN DISRAELI. Fcp. 8vo. 5s.

BOWDLER'S FAMILY SHAKSPEARE, cheaper Genuine Edition, complete in 1 vol. large type, with 36 Woodcut Illustrations, price 14s. or with the same ILLUSTRATIONS, in 6 pocket vols. 5s. each.

An ENGLISH TRAGEDY; Mary Stuart, from SCHILLER; and Mdlle. De Belle Isle, from A. DUMAS,—each a Play in 5 Acts, by FRANCES ANNE KEMBLE. Post 8vo. 12s.

Rural Sports, &c.

ENCYCLOPÆDIA of RURAL SPORTS; a complete Account, Historical, Practical, and Descriptive, of Hunting, Shooting, Fishing, Racing, &c. By D. P. BLAINE. With above 600 Woodcuts (20 from Designs by JOHN LEECH). 8vo. 42s.

COL. HAWKER'S INSTRUCTIONS to YOUNG SPORTSMEN in all that relates to Guns and Shooting. Revised by the Author's SON. Square crown 8vo. with Illustrations, 18s.

NOTES on RIFLE SHOOTING. By Captain HEATON, Adjutant of the Third Manchester Rifle Volunteer Corps. Fcp. 8vo. 2s. 6d.

The **DEAD SHOT**, or Sportsman's Complete Guide; a Treatise on the Use of the Gun, Dog-breaking, Pigeon-shooting, &c. By MARKSMAN. Fcp. 8vo. with Plates, 5s.

The **CHASE of the WILD RED DEER in DEVON and SOMERSET.** By C. P. COLLYNS. With Map and Illustrations. Square crown 8vo. 16s.

The **FLY-FISHER'S ENTOMOLOGY.** By ALFRED RONALDS. With coloured Representations of the Natural and Artificial Insect. 6th Edition; with 20 coloured Plates. 8vo. 14s.

HANDBOOK of ANGLING: Teaching Fly-fishing, Trolling, Bottom-fishing, Salmon-fishing; with the Natural History of River Fish, and the best modes of Catching them. By EPHEMERA. Fcp. Woodcuts, 5s.

The **CRICKET FIELD**; or, the History and the Science of the Game of Cricket. By J. PYCROFT, B.A. Trin. Coll. Oxon. 4th Edition. Fcp. 8vo. 5s.

The **Cricket Tutor**; a Treatise exclusively Practical. By the same. 18mo. 1s.

The **HORSE'S FOOT, and HOW to KEEP IT SOUND.** By W. MILES, Esq. 9th Edition, with Illustrations. Imp. 8vo. 12s. 6d.

A Plain Treatise on Horse-Shoeing. By the same Author. Post 8vo. with Illustrations, 2s.

General Remarks on Stables, and Examples of Stable Fittings. By the same. Imp. 8vo. with 13 Plates, 15s.

Remarks on Horses' Teeth, adapted to Purchasers. By the same Author. Crown 8vo. 1s. 6d.

The **HORSE**: with a Treatise on Draught. By WILLIAM YOUATT. New Edition, revised and enlarged. 8vo. with numerous Woodcuts, 10s. 6d.

The **Dog.** By the same Author. 8vo. with numerous Woodcuts, 6s.

The **DOG in HEALTH and DISEASE.** By STONEHENGE. With 70 Wood Engravings. Square crown 8vo. 15s.

The **Greyhound.** By the same. With many Illustrations. Square crown 8vo. 21s.

The **OX**; his Diseases and their Treatment: with an Essay on Parturition in the Cow. By J. R. DOBSON, M.R.C.V.S. Post 8vo. with Illustrations.
[*Just ready.*

Commerce, Navigation, and Mercantile Affairs.

The **LAW of NATIONS** Considered as Independent Political Communities. By TRAVERS TWISS, D.C.L. Regius Professor of Civil Law in the University of Oxford. 2 vols. 8vo. 30s. or separately, PART I. *Peace*, 12s. PART II. *War*, 18s.

A **DICTIONARY**, Practical, Theoretical, and Historical, of Commerce and Commercial Navigation. By J. R. M'CULLOCH, Esq. 8vo. with Maps and Plans, 50s.

The **STUDY** of **STEAM** and the **MARINE ENGINE**, for Young Sea Officers. By S. M. SAXBY, R.N. Post 8vo. with 87 Diagrams, 5s. 6d.

A **NAUTICAL DICTIONARY**, defining the Technical Language relative to the Building and Equipment of Sailing Vessels and Steamers, &c. By ARTHUR YOUNG. Second Edition; with Plates and 150 Woodcuts. 8vo. 18s.

A **MANUAL** for **NAVAL CADETS**. By J. M'NEIL BOYD, late Captain R.N. Third Edition; with 240 Woodcuts and 11 coloured Plates. Post 8vo. 12s. 6d.

⁎ Every Cadet in the Royal Navy is required by the Regulations of the Admiralty to have a copy of this work on his entry into the Navy.

Works of Utility and General Information.

MODERN COOKERY for **PRIVATE FAMILIES**, reduced to a System of Easy Practice in a Series of carefully-tested Receipts. By ELIZA ACTON. Newly revised and enlarged; with 8 Plates, Figures, and 150 Woodcuts. Fcp. 8vo. 7s. 6d.

On **FOOD** and its **DIGESTION**; an Introduction to Dietetics. By W. BRINTON, M.D. Physician to St. Thomas's Hospital, &c. With 48 Woodcuts. Post 8vo. 12s.

ADULTERATIONS DETECTED; or Plain Instructions for the Discovery of Frauds in Food and Medicine. By A. H. HASSALL, M.D. Crown 8vo. with Woodcuts, 17s. 6d.

The **VINE** and its **FRUIT**, in relation to the Production of Wine. By JAMES L. DENMAN. Crown 8vo. 8s. 6d.

WINE, the **VINE**, and the **CELLAR**. By THOMAS G. SHAW. With 28 Illustrations on Wood. 8vo. 16s.

A **PRACTICAL TREATISE** on **BREWING**; with Formulæ for Public Brewers, and Instructions for Private Families. By W. BLACK. 8vo. 10s. 6d.

SHORT WHIST; its Rise, Progress, and Laws; with the Laws of Piquet, Cassino, Ecarté, Cribbage, and Backgammon. By Major A. Fcp. 8vo. 3s.

HINTS on ETIQUETTE and the USAGES of SOCIETY; with a Glance at Bad Habits. Revised, with Additions, by a LADY of RANK. Fcp. 8vo. 2s. 6d.

The CABINET LAWYER; a Popular Digest of the Laws of England, Civil and Criminal. 19*th Edition*, extended by the Author; including the Acts of the Sessions 1862 and 1863. Fcp. 8vo. 10s. 6d.

The PHILOSOPHY of HEALTH; or, an Exposition of the Physiological and Sanitary Conditions conducive to Human Longevity and Happiness. By SOUTHWOOD SMITH, M.D. Eleventh Edition, revised and enlarged: with New Plates, 8vo.   [*Just ready.*

HINTS to MOTHERS on the MANAGEMENT of their HEALTH during the Period of Pregnancy and in the Lying-in Room. By T. BULL, M.D. Fcp. 8vo. 5s.

The Maternal Management of Children in Health and Disease. By the same Author. Fcp. 8vo. 5s.

NOTES on HOSPITALS. By FLORENCE NIGHTINGALE. Third Edition, enlarged; with 13 Plans. Post 4to. 18s.

C. M. WILLICH'S POPULAR TABLES for ascertaining the Value of Lifehold, Leasehold, and Church Property, Renewal Fines, &c.; the Public Funds; Annual Average Price and Interest on Consols from 1731 to 1861; Chemical, Geographical, Astronomical, Trigonometrical Tables, &c. Post 8vo. 10s.

THOMSON'S TABLES of INTEREST, at Three, Four, Four and a Half, and Five per Cent. from One Pound to Ten Thousand and from 1 to 365 Days. 12mo. 3s. 6d.

MAUNDER'S TREASURY of KNOWLEDGE and LIBRARY of Reference: comprising an English Dictionary and Grammar, a Universal Gazetteer, a Classical Dictionary, a Chronology, a Law Dictionary, a Synopsis of the Peerage, useful Tables, &c. Fcp. 8vo. 10s.

General and School Atlases.

An ELEMENTARY ATLAS of HISTORY and GEOGRAPHY, from the commencement of the Christian Era to the Present Time, in 16 coloured Maps, chronologically arranged, with illustrative Memoirs. By the Rev. J. S. BREWER, M.A. Royal 8vo. 12s. 6d.

SCHOOL ATLAS of PHYSICAL, POLITICAL, and COMMERCIAL GEOGRAPHY, in 17 full-coloured Maps, accompanied by descriptive Letterpress. By E. HUGHES, F.R.A.S. Royal 8vo. 10s. 6d.

BISHOP BUTLER'S ATLAS of ANCIENT GEOGRAPHY, in a Series of 24 full-coloured Maps, accompanied by a complete Accentuated Index. New Edition, corrected and enlarged. Royal 8vo. 12s.

BISHOP BUTLER'S ATLAS of MODERN GEOGRAPHY, in a Series of 33 full-coloured Maps, accompanied by a complete Alphabetical Index. New Edition, corrected and enlarged. Royal 8vo. 10s. 6d.

In consequence of the rapid advance of geographical discovery, and the many recent changes, through political causes, in the boundaries of various countries, it has been found necessary thoroughly to revise this long-established Atlas, and to add several new MAPS. New MAPS have been given of the following countries: *Palestine, Canada*, and the adjacent provinces of *New Brunswick, Nova Scotia*, and *Newfoundland*, the *American States* bordering on the Pacific, *Eastern Australia*, and *New Zealand*. In addition to these MAPS of *Western Australia* and *Tasmania* have been given in compartments; thus completing the revision of the MAP of *Australasia*, rendered necessary by the rising importance of our Australasian possessions. In the MAP of *Europe, Iceland* has also been re-drawn, and the new boundaries of *France, Italy*, and *Austria* represented. The MAPS of the three last-named countries have been carefully revised. The MAP of *Switzerland* has been wholly re-drawn, showing more accurately the physical features of the country. *Africa* has been carefully compared with the discoveries of LIVINGSTONE, BURTON, SPEKE, BARTH, and other explorers. The number of MAPS is thus raised from Thirty to Thirty-three. An entirely new INDEX has been constructed; and the price of the work has been reduced from 12s. to Half-a-Guinea. The present edition, therefore, will be found much superior to former ones; and the Publishers feel assured that it will maintain the character which this work has so long enjoyed as a popular and comprehensive School Atlas.

MIDDLE-CLASS ATLAS of GENERAL GEOGRAPHY, in a Series of 29 full-coloured Maps, containing the most recent Territorial Changes and Discoveries. By WALTER M'LEOD, F.R.G.S. 4to. 5s.

PHYSICAL ATLAS of GREAT BRITAIN and IRELAND; comprising 30 full-coloured Maps, with illustrative Letterpress, forming a Concise Synopsis of British Physical Geography. By WALTER M'LEOD, F.R.G.S. Fcp. 4to. 7s. 6d.

Periodical Publications.

The **EDINBURGH REVIEW**, or **CRITICAL JOURNAL**, published Quarterly in January, April, July, and October. 8vo. price 6s. each No.

The **GEOLOGICAL MAGAZINE**, or Monthly Journal of Geology, edited by T. RUPERT JONES, F.G.S. assisted by HENRY WOODWARD, F.G.S. 8vo. price 1s. 6d. each No.

FRASER'S MAGAZINE for **TOWN** and **COUNTRY**, published on the 1st of each Month. 8vo. price 2s. 6d. each No.

The **ALPINE JOURNAL**: a Record of Mountain Adventure and Scientific Observation. By Members of the Alpine Club. Edited by H. B. GEORGE, M.A. Published Quarterly, May 31, Aug. 31, Nov. 30, Feb. 28. 8vo. price 1s. 6d. each No.

INDEX.

Acton's Modern Cookery 26
Afternoon of Life 20
Alcock's Residence in Japan 21
Alpine Guide (The) 22
——— Journal (The) 28
Afjohn's Manual of the Metalloids 11
Arago's Biographies of Scientific Men 5
——— Popular Astronomy 10
——— Meteorological Essays 10
Arnold's Manual of English Literature 7
Arnott's Elements of Physics 11
Atherstone Priory 23
Atkinson's Papinian 5
Autumn Holiday of a Country Parson 8
Ayre's Treasury of Bible Knowledge 18

Bacon's Essays, by Whately 5
——— Life and Letters, by Spedding 3
——— Works, by Ellis Spedding, and Heath 5
Bain on the Emotions and Will 9
——— on the Senses and Intellect 9
——— on the Study of Character 9
Baines's Explorations in S. W. Africa 21
Ball's Guide to the Central Alps 22
——— Guide to the Western Alps 22
Bayldon's Rents and Tillages 17
Berlepsch's Life and Nature in the Alps .. 12
Black's Treatise on Brewing 26
Blackley and Friedlander's German and English Dictionary 8
Blaine's Rural Sports 24
Blight's Week at the Land's End 22
Bourne's Catechism of the Steam Engine .. 16
——— Treatise on the Steam Engine 16
Bowdler's Family Shakspeare 24
Boyd's Manual for Naval Cadets 26
Bramley-Moore's Six Sisters of the Valleys 23
Brande's Dictionary of Science, Literature, and Art 13
Bray's (C.) Education of the Feelings 9
——— Philosophy of Necessity 9
——— (Mrs.) British Empire 10
Brewer's Atlas of History and Geography .. 27
Brinton on Food and Digestion 26
Bristow's Glossary of Mineralogy 11
Brodie's (Sir C. B.) Psychological Inquiries 9
——— Works 14
Brown's Demonstrations of Microscopic Anatomy 14
Browne's Exposition of the 39 Articles ... 17
——— Pentateuch and Elohistic Psalms 17
Buckle's History of Civilization 2
Bull's Hints to Mothers 27
——— Maternal Management of Children ... 27

Bunsen's Analecta Ante-Nicæna 19
——— Ancient Egypt 3
——— Hippolytus and his Age 19
——— Philosophy of Universal History 19
Bunyan's Pilgrim's Progress, illustrated by Bennett 15
Burke's Vicissitudes of Families 4
Butler's Atlas of Ancient Geography 27
——— Modern Geography 28

Cabinet Lawyer 27
Calvert's Wife's Manual 20
Cats and Farlie's Moral Emblems 15
Chorale Book for England 21
Colenso (Bishop) on Pentateuch and Book of Joshua 18
Collyns on Stag-Hunting in Devon and Somerset 25
Commonplace Philosopher in Town and Country 8
Companions of my Solitude 8
Conington's Handbook of Chemical Analysis 13
Contanseau's Pocket French and English Dictionary 7
——— Practical ditto 7
Conybeare and Howson's Life and Epistles of St. Paul 19
Copland's Dictionary of Practical Medicine 14
——— Abridgment of ditto 14
Cotton's Introduction to Confirmation 19
Cox's Tales of the Great Persian War 2
——— Tales from Greek Mythology 23
——— Tales of the Gods and Heroes 23
——— Tales of Thebes and Argos 23
Cresy's Encyclopædia of Civil Engineering 16
Crowe's History of France 2

D'Aubigné's History of the Reformation in the time of Calvin 2
Dead Shot (The), by Marksman 25
De la Rive's Treatise on Electricity 11
Denman's Vine and its Fruit 26
De Tocqueville's Democracy in America ... 2
Diaries of a Lady of Quality 4
Disraeli's Revolutionary Epick 24
Dixon's Fasti Eboracenses 4
Dobson on the Ox 25
Döllinger's Introduction to History of Christianity 19
Dove's Law of Storms 10
Doyle's Chronicle of England 3

Edinburgh Review (The) 26
Ellice, a Tale 23
ELLICOTT'S Broad and Narrow Way 18
———— Commentary on Ephesians 18
———— Destiny of the Creature 18
———— Lectures on Life of Christ 18
———— Commentary on Galatians 18
———————————— Pastoral Epist... 18
———————————— Philippians, &c.. 18
———————————— Thessalonians ... 18
Essays and Reviews 19
Essays on Religion and Literature, edited by MANNING 19
Essays written in the Intervals of Business .. 8

FAIRBAIRN'S Application of Cast and Wrought Iron to Building 16
———— Information for Engineers... 16
———— Treatise on Mills & Millwork 16
First Friendship 22
FITZ ROY'S Weather Book 10
FORSTER'S Life of Sir John Eliot 3
FOWLER'S Collieries and Colliers 17
Fraser's Magazine 28
FRESHFIELD'S Alpine Byways 22
———— Tour in the Grisons 22
Friends in Council 6
From Matter to Spirit 8
FROUDE'S History of England 1

GARRATT'S Marvels and Mysteries of Instinct 12
Geological Magazine 11, 26
GILBERT and CHURCHILL'S Dolomite Mountains 22
GOODEVE'S Elements of Mechanism 16
GOYLE'S Questions on BROWNE'S Exposition of the 39 Articles 17
GRAY'S Anatomy 14
GREENE'S Manual of Coelenterata 11
———— Manual of Protozoa 11
GROVE on Correlation of Physical Forces .. 11
Gryll Grange 22
GWILT'S Encyclopædia of Architecture 15

Handbook of Angling, by EPHEMERA 26
HARTWIG'S Sea and its Living Wonders.... 12
———— Tropical World 12
HASSALL'S Adulterations Detected 26
———— British Freshwater Algæ ... 12
HAWKER'S Instructions to Young Sportsmen 25
HEATON'S Notes on Rifle Shooting 25
HELPS'S Spanish Conquest in America 2
HERSCHEL'S Essays from the Edinburgh and Quarterly Reviews 13
———— Outlines of Astronomy 9
HEWITT on the Diseases of Women 13
HINCHLIFF'S South American Sketches 21
HIND'S Canadian Exploring Expeditions ... 21
———— Explorations in Labrador ... 21
Hints on Etiquette 27
HOLLAND'S Chapters on Mental Physiology . 6
———— Essays on Scientific Subjects... 13
———— Medical Notes and Reflections.. 15
HOLMES'S System of Surgery 14
HOOKER and WALKER-ARNOTT'S British Flora 12
HOOPER'S Medical Dictionary 15
HORNE'S Introduction to the Scriptures ... 16
———— Compendium of ditto 16
HOSKYNS' Talpa 17
HOWITT'S History of the Supernatural 8
———— Rural Life of England 22
———— Visits to Remarkable Places.. 22

HOWSON'S Hulsean Lectures on St. Paul.... 18
HUGHES'S (E.) Atlas of Physical, Political and Commercial Geography............. 27
———— (W.) Geography of British History 10
———— Manual of Geography 10
HULLAH'S History of Modern Music........ 3
Hymns from *Lyra Germanica*.............. 20

INGELOW'S Poems........................... 24

JAMESON'S Legends of the Saints and Martyrs.................................. 15
———— Legends of the Madonna..... 15
———— Legends of the Monastic Orders 15
JAMESON and EASTLAKE'S History of Our Lord 15
JOHNS'S Home Walks and Holiday Rambles 12
JOHNSON'S Patentee's Manual 16
———— Practical Draughtsman...... 16
JOHNSTON'S Gazetteer, or Geographical Dictionary 10
JONES'S Christianity and Common Sense.... 9

KALISCH'S Commentary on the Old Testament 7
———— Hebrew Grammar............ 7
KEMBLE'S Plays 24
KENNEDY'S Hymnologia Christiana 20
KIRBY and SPENCE'S Entomology 12

Lady's Tour Round Monte Rosa 22
LANDON'S (L. E. L.) Poetical Works........ 24
Late Laurels 22
LATHAM'S Comparative Philology........... 6
———— English Dictionary 6
———— Handbook of the English Language 6
———— Work on the English Language 6
Leisure Hours in Town 8
LEWES'S Biographical History of Philosophy 3
LEWIS on the Astronomy of the Ancients .. 6
———— on the Credibility of Early Roman History 6
———— Dialogue on Government 6
———— on Egyptological Method..... 6
———— Essays on Administrations ... 6
———— Fables of BABRIUS 6
———— on Foreign Jurisdiction 6
———— on Irish Disturbances 6
———— on Observation and Reasoning in Politics 6
———— on Political Terms 6
———— on the Romance Languages ... 6
LIDDELL and SCOTT'S Greek-English Lexicon 7
———— Abridged ditto 7
LINDLEY and MOORE'S Treasury of Botany 13
LISTER'S Physico-Prophetical Essays 10
LONGMAN'S Lectures on the History of England 2
LOUDON'S Encyclopædia of Agriculture.... 17
———————————— Cottage, Farm, and Villa Architecture 17
———————————— Gardening..... 17
———————————— Plants...... 12
———————————— Trees & Shrubs 12
LOWNDES'S Engineer's Handbook 16
Lyra Domestica 21
———— Eucharistica 20
———— Germanica 15, 20
———— Messianica 20
———— Mystica 20
———— Sacra 20

NEW WORKS PUBLISHED BY LONGMAN AND CO. 31

MACAULAY's (Lord) Essays 3
———————— History of England 1
———————— Lays of Ancient Rome 24
———————— Miscellaneous Writings 8
———————— Speeches 6
———————— Speeches on Parliamentary Reform 6
MACBRAIR's Africans at Home 10
MACDOUGALL's Theory of War............ 16
McLEOD's Middle-Class Atlas of General Geography 28
———————— Physical Atlas of Great Britain and Ireland 28
McCULLOCH's Dictionary of Commerce 26
———————— Geographical Dictionary...... 10
MAGUIRE's Life of Father Mathew.......... 4
———————— Rome and its Rulers........... 4
MALING's Indoor Gardener 12
Maps from Peaks, Passes, and Glaciers 16
MARSHALL's History of Christian Missions . 3
MASSEY's History of England 1
MAUNDER's Biographical Treasury ———— 5
———————— Geographical Treasury 10
———————— Historical Treasury 3
———————— Scientific and Literary Treasury 13
———————— Treasury of Knowledge 27
———————— Treasury of Natural History .. 12
MAURY's Physical Geography 10
MAY's Constitutional History of England.. 1
MELVILLE's Digby Grand................... 23
———————— General Bounce 23
———————— Gladiators 23
———————— Good for Nothing ———————— 23
———————— Holmby House 23
———————— Interpreter 23
———————— Kate Coventry 23
———————— Queen's Maries............... 23
MENDELSSOHN's Letters.................... 4
MENTEITH's Windsor Great Park........... 17
MERIVALE's (H.) Colonisation and Colonies 10
———————— (C.) Fall of the Roman Republic 2
———————— Romans under the Empire .. 2
MERYON's History of Medicine.............. 3
MILES on Horse's Foot 25
———— on Horses' Teeth 25
———— on Horse Shoeing.................. 25
———— on Stables 25
MILL on Liberty 5
——— on Representative Government 5
——— on Utilitarianism................... 5
MILL's Dissertations and Discussions 5
———— Political Economy 5
———— System of Logic 5
MILLER's Elements of Chemistry............ 13
MONSELL's Spiritual Songs 20
MONTAGU's Experiments in Church and State.. 18
MONTGOMERY on the Signs and Symptoms of Pregnancy................................. 18
MOORE's Irish Melodies.................... 24
———————— Lalla Rookh 24
———————— Memoirs, Journal, and Correspondence 4
———————— Poetical Works............... 24
MORELL's Elements of Psychology 9
———————— Mental Philosophy 9
Morning Clouds 20
MORTON's Handbook of Dairy Husbandry.. 17
———————— Farm Labour 17
———————— Prince Consort's Farms..... 17
MOSHEIM's Ecclesiastical History 19
MÜLLER's (Max) Lectures on the Science of Language 7
———————— (K. O.) Literature of Ancient Greece 2
MURCHISON on Continued Fevers.......... 14
MURE's Language and Literature of Greece 2

New Testament Illustrated with Wood Engravings from the Old Masters............ 15
NEWMAN's Apologia pro Vita Suâ 3
NIGHTINGALE's Notes on Hospitals........ 27

ODLING's Course of Practical Chemistry 13
———————— Manual of Chemistry 13
ORMSBY's Rambles in Algeria and Tunis... 21
OWEN's Comparative Anatomy and Physiology of Vertebrate Animals 11

PACKE's Guide to the Pyrenees ————— 22
PAGET's Lectures on Surgical Pathology.. 14

PARKER's (Theodore) Life, by WEISS........ 4
Peaks, Passes, and Glaciers, 2 Series 21
PEREIRA's Elements of Materia Medica.... 15
———————— Manual of Materia Medica ... 15
PERKINS's Tuscan Sculptors 16
PHILLIPS's Guide to Geology 11
———————— Introduction to Mineralogy ... 11
PIESSE's Art of Perfumery 17
———————— Chemical, Natural, and Physical Magic 17
———————— Laboratory of Chemical Wonders 17
Playtime with the Poets 24
Practical Mechanic's Journal 16
PRESCOTT's Scripture Difficulties 18
Problems in Human Nature................. 20
PYCROFT's Course of English Reading..... 7
———————— Cricket Field 25
———————— Cricket Tutor 25

Recreations of a Country Parson, SECOND SERIES 8
RIDDLE's Diamond Latin-English Dictionary 7
RIVERS's Rose Amateur's Guide............ 12
ROGERS's Correspondence of Greyson 9
———————— Eclipse of Faith 9
———————— Defence of ditto 9
———————— Essays from the Edinburgh Review 9
———————— Fulleriana 9
———————— Reason and Faith........... 9
ROGET's Thesaurus of English Words and Phrases 7
RONALDS's Fly-Fisher's Entomology 25
ROWTON's Debater.......................... 7

NEW WORKS PUBLISHED BY LONGMAN AND CO.

Saxby's Study of Steam 26
———— Weather System 10
Scott's Handbook of Volumetrical Analysis 13
Scrope on Volcanos 11
Senior's Biographical Sketches 5
———— Essays on Fiction 23
Sewell's Amy Herbert 23
———— Ancient History 3
———— Cleve Hall 23
———— Earl's Daughter 23
———— Experience of Life 23
———— Gertrude 23
———— Glimpse of the World 23
———— History of the Early Church ... 3
———— Ivors 23
———— Katharine Ashton 23
———— Laneton Parsonage 23
———— Margaret Percival 23
———— Night Lessons from Scripture .. 20
———— Passing Thoughts on Religion .. 20
———— Preparation for Communion 20
———— Readings for Confirmation 20
———— Readings for Lent 20
———— Self-Examination before Confirmation 20
———— Stories and Tales 23
———— Thoughts for the Holy Week ... 20
———— Ursula 23
Shaw's Work on Wine 26
Snedden's Elements of Logic 6
Short Whist 26
Sieveking's (Amelia) Life, by Winkworth .. 4
Smith's (Southwood) Philosophy of Health . 27
———— (J.) Voyage and Shipwreck of St. Paul 19
———— (G.) Wesleyan Methodism 3
———— (Sydney) Memoir and Letters ... 4
———— Miscellaneous Works 8
———— Sketches of Moral Philosophy . 8
———— Wit and Wisdom 8
Southey's (Doctor) 7
———— Poetical Works 24
Stebbing's Analysis of Mill's Logic 6
Stephenson's (R.) Life by Jeaffreson and Pole 3
Stephen's Essays in Ecclesiastical Biography 5
———— Lectures on the History of France 2
Stonehenge on the Dog 25
———— on the Greyhound 25
Strickland's Queens of England 1

Taylor's (Jeremy) Works, edited by Eden .. 19
Tennent's Ceylon 12
———— Natural History of Ceylon 12
———— Story of the Guns 16
Thalatta 23

Theologia Germanica 19
Thirlwall's History of Greece 2
Thomson's (Archbishop) Laws of Thought .. 6
———— (J.) Tables of Interest 27
Tilley's Eastern Europe and Western Asia . 21
Todd's Cyclopædia of Anatomy and Physiology 14
———— and Bowman's Anatomy and Physiology of Man 14
Trollope's Barchester Towers 23
———— Warden 23
Twiss's Law of Nations 26
Tyndall's Lectures on Heat 11
———— Mountaineering in 1861 22

Ure's Dictionary of Arts, Manufactures, and Mines 16

Vander Hoeven's Handbook of Zoology 11
Vaughan's (R.) Revolutions in English History 1
———— (R. A.) Hours with the Mystics 9
Warburton's Life, by Watson 4
Warter's Last of the Old Squires 23
Watson's Principles and Practice of Physic 14
Watts's Dictionary of Chemistry 13
Webb's Celestial Objects for Common Telescopes 10
Webster & Wilkinson's Greek Testament ... 18
Weld's Last Winter in Rome 21
Wellington's Life, by Brialmont and Gleig 4
———— by Gleig 4
Wesley's Life, by Southey 4
West on the Diseases of Infancy and Childhood 13
Whately's English Synonymes 5
———— Logic 5
———— Remains 5
———— Rhetoric 5
Whewell's History of the Inductive Sciences 2
White and Riddle's Latin-English Dictionary 7
Wilberforce (W.) Recollections of, by Harford 4
Willich's Popular Tables 27
Wilson's Bryologia Britannica 12
Wood's Homes without Hands 11
Woodward's Historical and Chronological Encyclopædia 3

Yonge's English-Greek Lexicon 7
———— Abridged ditto 7
Young's Nautical Dictionary 26
Youatt on the Dog 25
———— on the Horse 25

SPOTTISWOODE AND CO., PRINTERS, NEW-STREET SQUARE, LONDON

www.ingramcontent.com/pod-product-compliance
Lightning Source LLC
Chambersburg PA
CBHW030311170426
43202CB00009B/961